Essentials of Math

with Business Applications

Alvey • Johnson Nelson

Sixth Edition

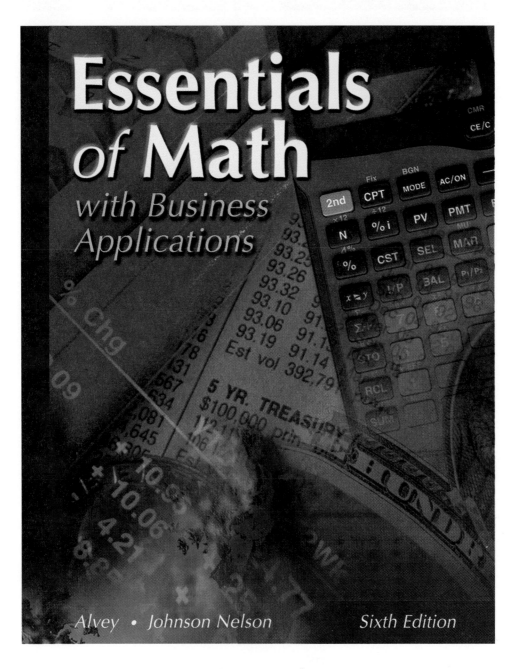

Essentials
of Math
with Business Applications

Alvey • Johnson Nelson   Sixth Edition

C. George Alvey
Professor Emeritus of Business Administration
County College of Morris
Randolph, New Jersey

Marceda Johnson Nelson
Kaw Area Technical School
Topeka, Kansas

Glencoe
McGraw-Hill

New York, New York   Columbus, Ohio   Woodland Hills, California   Peoria, Illinois

Cover Photo: Index Stock

All material from USA TODAY. Copyright 1998, USA TODAY. Reprinted with permission.

For permission to use copyrighted material, grateful
acknowledgment is made to the copyright holders on pp. 497,
which is hereby made part of this copyright page.

Essentials of Math with Business Applications   ISBN 0-02-643476-8
Instructor's Annotated Edition   ISBN 0-02-643478-4
Instructor's Resource Book   ISBN 0-02-643477-6

Glencoe/McGraw-Hill

A Division of The **McGraw·Hill** Companies

Essentials of Math with Business Applications, Sixth Edition
Copyright © 2001 by Glencoe/McGraw-Hill. All rights reserved. Copyright © 1995,
1989, 1984, 1976, 1968 by The McGraw-Hill Companies, Inc. All rights reserved.
Except as permitted under the United States Copyright Act, no part of this publication may be reproduced or distributed in any form or by any means, or stored in a
database or retrieval system, without the prior written permission of the publisher.

Send all inquiries to:
Glencoe/McGraw-Hill
936 Eastwind Drive
Westerville, OH 43081

ISBN 0-02-643476-8

Printed in the United States of America.

1 2 3 4 5 6 7 8 9 10 11 12 13 14 15   073   08 07 06 05 04 03 02 01 00 99

CONTENTS

UNIT 5 PERCENTS 222

SKILLBUILDER

UNIT 6 BUSINESS AND CONSUMER MATH 302

SKILLBUILDER

UNIT 7 INTEREST AND DISCOUNTS 368

SKILLBUILDER

PREFACE

The sixth edition of *Essentials of Math With Business Applications* is a comprehensive revision of the *Essentials of Business Math,* fifth edition. While retaining the emphasis on preparing students to be successful in today's workforce and in everyday life, the authors have focused attention on applications in a business environment and on reinforcement by increasing the number of problems and business applications. *Essentials of Math With Business Applications* is designed to be user friendly for students and instructors.

Purpose

The goal of this program is to assist students in achieving the Learning Objectives stated for each Skillbuilder by:

1. Reviewing the fundamentals of addition, subtraction, multiplication, and division through practice with computations involving whole numbers, decimals, fractions, and percents.

2. Relating the students' computational skills to typical business transactions involving interest, discount, payroll, depreciation, retail selling, and checking accounts.

3. Developing students' ability to use shortcut methods and to work with speed and accuracy.

4. Developing students' awareness and usage of the latest business technologies.

5. Developing students' study skills through Student Success Hints.

This text has been designed so that it can be used by students who are working on an individual basis as well as by students who are being instructed as a group.

Organization

The text is organized in seven units with seven to eleven Skillbuilders per unit. Each Skillbuilder states the expected Learning Objectives and introduces a new mathematical skill using *easy-to-read* and *easy-to-comprehend* explanations. Skillbuilders contain Math Tips, Calculator Tips, and Student Success Hints. Many Skillbuilders also include Business Applications for the skill introduced.

The Learning Outcomes and explanations are followed by exercises that include problems and application activities. The problems are arranged to progress from the simple to the more complex. Since the ultimate goal is to apply these skills in life and in the workforce, Business Applications are included in most Skillbuilders.

Because of the stated objectives, the clear and concise explanations and the integration of Business Applications, this text is flexible and can be used in individual or group situations. In addition, the text benefits students at multiple mathematical skill levels. The text has been designed to help students make successful school-to-work transitions.

Textbook Features

- **Four-Color Format.** A four-color format allows functional use of color for emphasis and in the presentation of invoices and other types of business forms.

- **Illustrations.** This text uses color illustrations to lend a realistic quality to business application material.

- **Skillbuilder Design.** The text has an open design, providing ample space for students to work problems.

- **New Activity Page(s).** An Activity Page(s) appears at the end of each Skillbuilder. Each Activity Page contains a USA TODAY Graph Activity, Challenge Activity, and an Internet Activity. The Internet Activity will enhance students' mastery of today's technology.

- **Margin Features.** Each Skillbuilder contains a Math Tip, Calculator Tip, and Student Success Hint.

- **More Problems.** This edition includes more problem material to enhance students' mastery of each new skill.

- **Competency-Based Skillbuilder.** Learning Objectives introduce each Skillbuilder.

- **Self-Check Problems.** These problems follow a worked-out example and provide immediate reinforcement for the student.

Instructor's Support Components

- The sixth edition *Instructor's Annotated Edition* features include:

  - Full student pages with annotations and worked-out solutions shown in place.

  - Correlation of the textbook to the SCANS competencies for easy reference.

  - Assignment Guide that provides recommended goals for student progress.

  - Cooperative Learning overview of suggestions on how to use cooperative learning activities with this program and how to integrate effective learning strategies in the classroom.

  - New User's Guide for the ExamView Test Generator and PowerPoint Slides Software.

- A new *Instructor's Resource Book* is also available. It contains:

  - *Practice Masters* that provide extra practice for each Skillbuilder.

  - *Metric Applications* provides a concise guide to metric units and conversions with problems and exercises for student practice.

  - *Civil Service Test,* a sample of the type of math problems to be expected on typical civil service and employment tests.

  - *Progress Reports* that can be duplicated for the students to use to mark their progress toward goals.

  - *GED Matrix* correlates the text to topics covered in the GED math test and provides practical suggestions for preparing for the test.

  - *Pretests* are available for each of the seven units. They provide a way for students to know their level of skill and knowledge before beginning a unit.

- New to the sixth edition is Software on CD-ROM. It contains:
  - A Test Generator that allows you to create a test in less than five minutes, select test questions, or edit questions, or add your own questions.
  - PowerPoint Slides are also included that contain examples and bulleted lists for various concepts.

Reviewers

Angelina Aysen
Louisiana Technical College
Thibodaux, Louisiana

Kim Belden
Daytona Beach Community College
Daytona Beach, Florida

Stephen J. Galambos
Philadelphia Job Corps Center
Philadelphia, Pennsylvania

Elizabeth Keller
South Hills School of Business &
Technology
State College, Pennsylvania

Carolyn Lewis
Lawrence Career College
Lawrence, Kansas

Colleen Morris
Chaparral College
Tucson, Arizona

James A. Page
Mountainview College
Dallas, Texas

Brenda A. B. Smith
State Technical Institute at Memphis
Memphis, Tennessee

Dr. John D. Walker
International Business College
Lubbock, Texas

Acknowledgments

The authors are grateful to the following contributors and business professionals for their contributions during the development of this text: Deann Blankenship, Account Clerk Instructor; Jeanette C. Stauffer, J. D., Legal Secretary Advisor/Instructor; and Christine Huntsman, Interior Design Instructor, Kaw Area Technical School, Topeka, KS; Arlene Brittain, Real Estate Agent and Retired Reading Specialist, Woodward, OK; Sue Schlegel, Director of Professional, Technical, and Vocational Education, Pasco School District, Pasco, WA; and Kathy Markham, Assistant to Secretary of Appointments for Governor of Kansas, Topeka, KS.

Essentials of Math

with Business Applications

Alvey • *Johnson Nelson*

Sixth Edition

UNIT 1

WHOLE NUMBERS
Addition and Subtraction

In this unit, you will study the following Skillbuilders:

Numbers dominate the business world. We use numbers to prepare payrolls, keep track of profit or loss, study industry trends, and so on. This pie chart shows a college's budget for publishing its student newspaper. What is the total budget?

In this unit we study addition and subtraction of whole numbers and money, including ways to make computation easier and to check our answers.

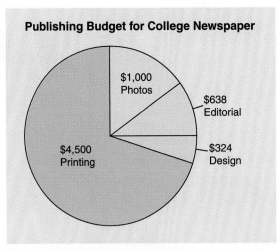

Publishing Budget for College Newspaper

- $1,000 Photos
- $638 Editorial
- $4,500 Printing
- $324 Design

MATH CONNECTIONS

Chef

Angela is the head chef at a restaurant that also caters parties. Although Angela's main job is to prepare food, she needs many of the same qualities as a small business owner. She manages the kitchen staff, which requires interviewing skills, as well as good interpersonal skills. Angela must also plan a strategy for the preparation of the food, so that each dish is completed on schedule. In addition, she must be able to forecast the food requirements; too little food will end with disappointed customers and too much food will result in waste. In order to make the necessary food projections, she must be able to estimate, add, and subtract in order to decide on food purchasing amounts.

Math Application

Angela is in charge of planning the "Happy Holidays" dinner for a large insurance company, which means making arrangements for approximately 700 people. The menu for the dinner includes shrimp cocktail and seafood thermidor. The shrimp cocktail requires 350 pounds of shrimp, and the seafood thermidor calls for 225 pounds of shrimp. How much shrimp must the chef buy in order to prepare the dinner?

(*a*) Add the necessary amounts to compute the total shrimp required.

Suppose Angela has 110 pounds of shrimp that she purchased and froze last week when it was on sale. How much shrimp must be purchased if the frozen shrimp is used?

(*b*) Subtract the number of frozen pounds from the total you found in (*a*).

Critical Thinking Problem

When Angela arrives at the seafood mart to purchase the shrimp, they only have 400 pounds left. What are some alternatives that she might use to solve this shortage of shrimp?

SKILLBUILDER 1.1

Reading and Writing Whole Numbers Using Words

Learning Objectives

• **Read and write numbers using words.**
• **Read word names of numbers and write numbers using digits.**

Reading and Writing Whole Numbers

Our number system is the **decimal,** or **base 10, system.** It uses the digits 1, 2, 3, 4, 5, 6, 7, 8, 9, and 0. The position of a digit indicates the place value of that digit.

The place value of each digit in a whole number is ten times greater than that of the place to its right. Thus, a digit in the tens place is worth ten times the same digit in the units place, and a digit in the hundreds place is worth ten times the same digit in the tens place.

In numbers greater than three places, commas are used to separate every three digits. Each group of three digits (except the units group) is read as a three-digit number followed by the name of the group to which it belongs.

MATH TIP

Note that a hyphen is used when compound numbers less than 100 are written in words. In other words, all numbers from twenty-one to ninety-nine that are made up of two number names are hyphenated.

EXAMPLE

Write 4,023,905 in words.

SOLUTION

Four million, twenty-three thousand, nine hundred five

Self-Check
1. Write 809 in words.

Self-Check Answer
Eight hundred nine

Problems

Read these numbers, and then write them in words.

1. 83 _____

2. 223 _____

3. 5,908 _____

4. 17,080 _____

5. 32,745 _____

6. 114,811 _____

7. 324,802 _____

8. 1,843,630 _____

9. 916 _____

10. 43 _____

11. 5,492 _____

12. 342,104,007 _____

13. 999 _____

14. 63,389 _____

15. 8,003 _____

16. 11,012 _____

17. 22 _____

18. 157 _____

Writing Whole Numbers Using Digits

We can use a place-value chart to help write numbers using digits. The following chart shows the number thirteen billion, seven hundred fifty million, two hundred forty-five thousand, five hundred seventy-one.

WHOLE NUMBERS														
Billions Group				Millions Group				Thousands Group				Units Group		
Hundred Billions	Ten Billions	Billions	Comma	Hundred Millions	Ten Millions	Millions	Comma	Hundred Thousands	Ten Thousands	Thousands	Comma	Hundreds	Tens	Units
	1	3	,	7	5	0	,	2	4	5	,	5	7	1

EXAMPLE

Write twenty-two thousand, eighty-six using digits.

SOLUTION

22,086

 Self-Check
Write nine thousand, forty-three using digits.

Self-Check Answer
9,043

Problems

Read these numbers, and then write them using digits.

19. Seventy-one

20. Sixty-three

21. Two hundred forty-nine

22. Nine hundred seven

23. Sixteen thousand, two hundred three

24. One hundred thirty-three thousand, four hundred fifty-nine

25. Five million, four hundred fifty thousand, nine

26. Three thousand, four hundred seventy-six

27. Seven hundred fifty-seven

28. Ninety-five

29. Eight million, five thousand, nine

30. Five hundred nine thousand, nine hundred forty-eight

31. Six million, three thousand, eight

32. Thirty-nine

33. One thousand, four hundred thirty-eight

34. Sixty-six

35. Four thousand, fifty-three

Answers

19._____

20._____

21._____

22._____

23._____

24._____

25._____

26._____

27._____

28._____

29._____

30._____

31._____

32._____

33._____

34._____

35._____

Activities 1.1

Graph Activity

The chart below shows the total U.S. flag imports and exports (in dollars) and the leading country for each. Write each number in words.

_____

_____

_____

_____

American flag trade deficit

The USA imports more Stars and Stripes than it exports. Total U.S. flag imports and exports (in dollars) in 1997 and leading country for each:

Total U.S. flags imports — $710,200
Top source: Taiwan — $566,700
Total U.S. flag exports — $473,200
Top destination: Dominican Republic — $102,400

Source: Census Bureau   By Cindy Hall and Genevieve Lynn, USA TODAY

Challenge Activity

Look in a local newspaper at a car dealership's ad. Choose 5 cars. Write the selling price using word names of the prices.

Internet Activity

Fabric manufacturing makes use of whole numbers in many ways. Find out the numbers that everyone will be wearing this year at **http://www.maa.org/ mathland/mathtrek_7_20_98.html**

What other products, according to this site, can use mathematical designs?

SKILLBUILDER 1.2

Aligning Digits by Place Value

Learning Objective

- **Align a column of numbers according to place value.**

Aligning Digits

Because place value is the governing factor in any arithmetic computation, it is essential that numbers be aligned according to the place value of the digits. The units place should be aligned with the units place, the tens place with the tens place, and so on.

EXAMPLE

Align the numbers 300 and 29 so that they can be added.

SOLUTION

Incorrectly Aligned

```
  300
+ 29      Incorrect
  590     Answer
```

Correctly Aligned

```
  300
+ 29      Correct
  329     Answer
```

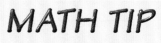

MATH TIP

Failure to write numbers clearly and to align them properly is one of the major reasons for errors in arithmetic computations.

Self-Check
Write the following numbers using digits in the columns at the right. Align them correctly both horizontally and vertically.

1. Four hundred three
2. Twenty-five
3. Two
4. Three thousand, twenty-nine
5. Fifty-two

Self-Check Answers

			4	0	3
				2	5
					2
		3	0	2	9
				5	2

Problems

Align the following groups of numbers according to place value. Copy them over so that digits of the same place value are aligned.

Answers

1. _____
   _____
   _____
   _____

2. _____
   _____
   _____
   _____

3. _____
   _____
   _____
   _____

4. _____
   _____
   _____
   _____

1. 472
   500
   6,843
   1,285

2. 2,705
   63
   287
   10,540

3. 9,453; 9,354; 36; 107

4. 10,057; 9,003; 124; 129,646

5. The groups of numbers below and on the next page are from different computer printouts. In the first group, the number zero is printed with a slash through it so it will not be confused with the uppercase letter "O." In the second group on the next page, the zero is printed with a dot in its center. Write these groups of numbers in words.

   a. 66665   5a. _____
      61026   _____
      35502   _____
      8265   _____
      95238   _____
      54952   _____

Student Success Hints

Ask Questions
In any learning situation, it is important to ask yourself and others questions for clarification and understanding.

b. $ 5 0 0   b. _____

99   _____

105   _____

109   _____

115   _____

123   _____

6. Write the following group of numbers using digits in a vertical column. Align digits of the same value.

    Four thousand, fifty-three
    Six hundred seven
    Eight thousand, eight hundred eight
    Seventeen thousand, thirty-two
    Sixty-four

Answers

6. _____

    _____

    _____

    _____

    _____

Write the following numbers using the digits in the columns at the right. Align them correctly both horizontally and vertically.

7. a. One hundred twelve

   b. Five thousand, fifty-two

   c. Twenty thousand, three hundred one

   d. Fifty-six

   e. Three hundred two

7. a.
   b.
   c.
   d.
   e.

8. a. Two hundred eighty-six

   b. Four thousand, five hundred ninety-seven

   c. Ninety-nine

   d. Nine thousand, two

   e. Eight hundred twenty-four

8. a.
   b.
   c.
   d.
   e.

Calculator Tip

To add and subtract money values, set the calculator in the "Add Mode" position if that function is available; if it is not, set the decimal selector to two.

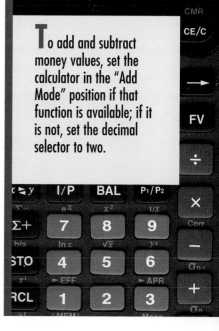

9. a. Seventy-six

   b. Four hundred eighty-nine

   c. Nine hundred ninety-nine

   d. Thirty-four

   e. Eighteen

9. a.
   b.
   c.
   d.
   e.

Activities 1.2

Graph Activity

Look at the graphic below. How many "hits" will Colorado State's tourism web site have in one year? List the place value of each digit in your answer.

Answers

_____

_____

_____

_____

_____

_____

Tourists on the Web

Tourism offices in the 50 states will spend an average $59,900 over the next 12 months developing and maintaining their Web sites. State tourism Web sites with the highest average "hits" per week:

206,000
Colorado

100,000
Ohio

175,000
North Carolina

120,000
Texas

200,000
California

Source: Travel Industry Association of America   By Cindy Hall and Quin Tian, USA TODAY

Challenge Activity

Look in the local newspaper at the real estate ads. Choose 5 homes. Write the prices in columns, placing the correct digits in the proper place value. List place values at the top of the columns.

Internet Activity

Calling all card sharks! Here's a game to practice your understanding of place values.

http://edweb.sdsu.edu/courses/edtec670/ Cardboard/card/n/NumberClub.html

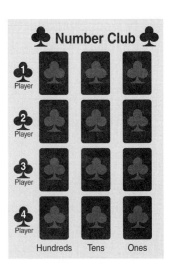

♣ Number Club ♣

1 Player
2 Player
3 Player
4 Player

Hundreds   Tens   Ones

SKILLBUILDER 1.3

Adding Whole Numbers

Learning Objective

• **Add two or more whole numbers.**

Adding Whole Numbers

In an addition problem, the numbers being added are called the **addends.**
The result of the computation is called the **total,** or **sum.** The order in
which the addends are added does not affect the sum.

$$2 + 4 = 6 \qquad 4 + 2 = 6$$

When adding a column of numbers that totals more than 10, it is necessary
to carry the first digit of that total to the next column of numbers.

EXAMPLE

Add: $939 + 147 + 265$.

SOLUTION

```
 1 2
 939
 147
+265
1,351
```

The total of the digits in the units place is 21. The *1* is written under
the column that is added, and the *2* is carried to the tens column. The
total of the digits in the tens column is 15. The *5* is written under the
tens column, and the *1* is carried to the next column to be added,
the hundreds column.

MATH TIP

*About one-half of the cal-
culations in business
involve addition. Therefore,
check addition by chang-
ing the order of the
addends.*

Self-Check
Add: $526 + 1,034 + 48 + 885$.

**Self-Check
Answer**
2,493

Problems

Add the numbers mentally. Then write the sums.

	a.	b.	c.	d.	e.	f.	g.	h.	i.	j.
1.	1	3	5	6	2	8	6	9	9	5
	7	9	7	4	6	7	3	9	5	5

	a.	b.	c.	d.	e.	f.	g.	h.	i.	j.
2.	3	6	6	3	1	7	1	2	5	3
	4	6	8	9	1	9	7	5	6	4

	a.	b.	c.	d.	e.	f.	g.	h.	i.	j.
3.	1	6	8	5	2	6	9	5	3	6
	4	2	7	8	3	2	3	6	2	6
	3	5	4	9	7	6	1	9	3	6

	a.	b.	c.	d.	e.	f.	g.	h.	i.	j.
4.	1	7	4	5	6	3	2	4	4	8
	5	4	1	8	5	9	4	3	4	7
	6	4	5	2	7	8	3	6	9	2
	8	2	3	4	4	7	5	8	6	7

Some of the following totals are wrong. Cross out any incorrect answer, and write the correct answer beside it.

	a.	b.	c.	d.	e.	f.	g.	h.	i.	j.
5.	9	7	9	5	3	7	4	9	6	2
	6	5	8	6	7	9	7	6	4	9
	8	2	4	4	9	9	6	7	4	9
	3	6	7	4	3	8	7	3	6	2
	26	20	27	19	22	33	34	25	20	22

	a.	b.	c.	d.	e.	f.	g.	h.	i.	j.
6.	17	10	47	70	53	38	45	85	22	31
	21	25	11	19	42	31	22	30	46	32
	41	34	20	20	41	10	51	53	11	66
	79	57	78	109	136	79	108	168	89	129

Name:_____ Date:_____

Add the numbers in the following problems.

	a.	b.	c.	d.	e.	f.	g.	h.	i.	j.
7.	89	132	342	45	425	321	1,543	243	2,023	34
	92	88	109	277	231	606	481	1,006	29	930
					456	195	7,328	751	871	256
									530	11

8. Find the total sales for each day of the week.

DAILY SALES BY REPRESENTATIVE

Representative	M	T	W	T	F
Richards	$ 765	$ 875	$1,254	$ 922	$1,135
Wolf	813	934	983	1,268	820
Bookner	906	1,007	834	785	955
Tobin	1,106	843	912	872	1,010
Wood	751	1,254	840	–0–	975
Total					

Student Success Hints

Organize for Success
- Organize time
  —Use a daily/ weekly schedule
  —Use a calendar system
- Organize materials and study space
  —3-ringed tabbed notebook
  —Zippered pocket for calculator, pencils, pens, etc.
  —Set aside a study space (area should be comfortable, uncluttered, and free from distractions)

9. Megan's Gifts sold 88 gift baskets in June, 109 gift baskets in July, and 72 gift baskets in August. How many gift baskets were sold in those three months?

10. During the first week of registration, Clinton Community College enrolled 438 students on Monday, 281 on Tuesday, 339 on Wednesday, 159 on Thursday, and 512 on Friday. How many students registered at Clinton Community College during the first week of registration?

Answers

9. _____

10. _____

Activities 1.3

Graph Activity

The graphic shows the number of women on active duty in each branch of the military. How many women are on active duty in the military?

Answer

_____

Women in the military

Former Defense secretary William Perry today receives the Margaret Chase Smith Award for advancing women in the military. It is named for the senator who wrote the 1948 law giving military women permanent status. Women on active duty:

Service (% total)	Women
Army (15%)	70,643
Navy (13%)	48,138
Air Force (18%)	65,285
Marine Corps (6%)	9,440
Coast Guard (10%)	3,288

Source: Women In Military Service for America Memorial

By Anne R. Carey and Elys A. McLean, USA TODAY

Challenge Activity

Make a list of everything you eat for one week's time. Record the amount of fat grams, fiber grams, and calorie intake for each day. At the end of the day add the fat grams, fiber grams, and the number of calories you consumed. On day seven add the total fat grams, fiber grams, and calories for the week. How many grams of fat, how many grams of fiber and how many calories did you consume?

Internet Activity

You and your best friend are going to see bears and whales in Alaska. Your wildlife excursions are piloted by Michelle, owner of Island Wings Air Service. To do both trips, what will the total cost be for both of you?

http://www.ktn.net/iwas/wlv.html

ALASKA

SKILLBUILDER 1.4

Developing Speed in Addition

Learning Objectives

- **Add whole numbers quickly by adding tens and then ones.**
- **Add whole numbers quickly by grouping sums of 10.**

Add Tens First

One method of adding two-digit numbers quickly is to mentally add all the tens digits and one of the units digits first. Then add the remaining units digits to the first total.

EXAMPLE

Add: $13 + 49 + 52$.

SOLUTION
To add $13 + 49 + 52$, first add $13 + 40 + 50 = 103$. Then add $103 + 9 = 112$ and $112 + 2 = 114$.

```
  13        13
  49        40
 +52       +50
         103 + 9 = 112
         112 + 2 = 114
```

Self-Check
Use the method described here to find $55 + 83$.

MATH TIP

It is helpful to be able to recognize three numbers that total 10. These combinations may appear in different sequences, for example, $2 + 3 + 5 = 10$, $3 + 5 + 2 = 10$, $3 + 2 + 5 = 10$, $5 + 2 + 3 = 10$, $5 + 3 + 2 = 10$, and $2 + 5 + 3 = 10$.

Self-Check Answer
138

Problems

Find the sum in each of the following problems. Use the method described on the previous page.

	a.	b.	c.	d.	e.	f.	g.	h.	i.	j.
1.	26	19	43	79	68	22	38	57	49	83
	57	86	76	44	87	16	41	35	86	71
						83	96	48	28	68
2.	49	42	19	39	68	75	48	62	41	31
	56	61	93	67	86	15	37	28	60	48
	62	98	42	85	40	53	82	74	39	99
	78	48	78	67	69	98	52	85	58	57
						84	99	97	75	89

Add By Grouping

Another way to develop speed in addition is to add using groups of two or more digits that total 10.

1	2	3	4	5	6	7	8	9
+9	+8	+7	+6	+5	+4	+3	+2	+1
10	10	10	10	10	10	10	10	10

First add any digits in a column that total 10. Then add this total to the remaining digits.

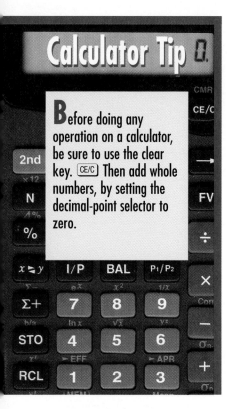

Student Success Hints

Visualize
Get a mental picture of what you are doing— recall past experiences or knowledge that brings you to the point of new learning.

Self-Check Answer
291

EXAMPLE

Add: 96 + 54 + 63 + 12 + 55.

SOLUTION

The digits connected by lines total 10.

$$
\begin{array}{r}
2 \\
96 \\
10 = \quad 54 \quad = 10 \\
63 \\
10 = \quad 12 \quad = 10 \\
+55 \\
\hline
280
\end{array}
$$

Note that 10 tens + 10 tens + 6 tens = 26 tens, or 260.
Then 260 + 10 + 10 = 280.

Self-Check
Add by using groups that total 10:
27 + 69 + 71 + 43 + 81.

Problems

Add the numbers in these problems by using groups that total 10.

	a.	b.	c.	d.	e.	f.	g.
3.	216	547	813	5,426	4,875	498	$928
	461	164	726	1,251	9,234	953	163
	628	432	652	4,943	1,768	691	42
	593	634	295	3,942	3,312	462	5
	836	522	134	5,627	6,241	312	675
						249	12
						527	415

4.	928	3,613	156	492	333	611	738
	286	9,149	904	816	777	994	327
	159	2,572	106	329	888	653	673
	462	8,426	936	817	222	457	437
	543	6,321					
	352	4,157					
	621	2,212					

Using either method described, find the sum of each of the following groups.

	a.	b.	c.	d.	e.
5.	427	506	468	683	492
	68	82	49	91	84
	127	104	106	121	114
	653	602	608	801	590
	371	360	346	294	361

	f.	g.	h.	i.	j.
	92	955	86	365	513
	742	91	256	49	47
	436	484	427	559	562
	328	379	258	382	432
	685	764	513	365	607

Activities 1.4

Graph Activity

Without using a calculator, calculate the total number of players and spouses that went back to college between 1992 and 1998.

Answer

_____

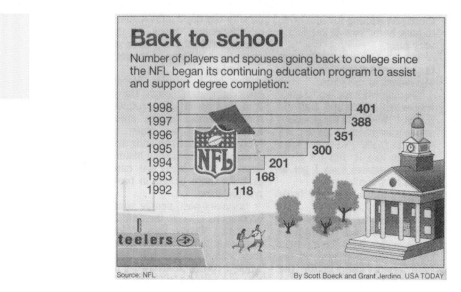

Back to school

Number of players and spouses going back to college since the NFL began its continuing education program to assist and support degree completion:

Year	Number
1998	401
1997	388
1996	351
1995	300
1994	201
1993	168
1992	118

Source: NFL   By Scott Boeck and Grant Jerding, USA TODAY

Challenge Activity

Keep a record of your car mileage each day for a week and add the total miles driven. (If you are a walker or runner keep a record of the miles walked/ran.) Calculate the total mileage.

Internet Activity

According to recent census reports, how much would a two-adult household spend per year on household operations, housekeeping supplies, and household furnishings and equipment? How much for dairy products, fruits, and vegetables?

ftp://ftp.bls.gov/pub/special.requests/ce/ standard/1996/cucomp.txt

SKILLBUILDER 1.5

Checking Addition

Learning Objective

• **Check the accuracy of addition.**

Checking Addition

Never assume that the addition of a column of figures is correct. If figures were copied from another source, check the copy against the original to be sure that the figures were copied accurately. Check the accuracy of addition done by hand by adding the same figures in the reverse order.

EXAMPLE

Add and check: $385 + 54 + 915 + 446$.

MATH TIP

Remember, the possibility of error is introduced when copying numbers from one document to another.

SOLUTION

$$1,800$$

	385	
	54	
Then add up.	915	First add down.
	446	
	1,800	

Self-Check
Add and check: $1,024 + 378 + 2,592 + 74$.

Self-Check Answer
4,068

Problems

Copy each of the following groups of figures in the space provided, find the total, and then check the total by adding in reverse order. Try using the speed methods presented in Skillbuilder 1.4.

1.

57			
36			
48			
24			

2.

386			
159			
514			
212			

3.
$6,321
246
2,179
1,532

4.
$54,943
7,156
63,041
6,119

5.
35
16
84
25
57

6.
457
123
316
286
934

7.
$15,726
3,168
9,003
213
1,124

8.
$18,182
717
4,111
13,155
2,435

9.
10,345
5,876
11,490
30,467
3,041

10.
6,098
23,086
10,137
9,437
897

11.
876
12,432
2,376
690
4,639

12.
$12,387
4,592
5,231
10,429
234

© Glencoe/McGraw-Hill

Business Applications 1.5

1. This record shows the number of bicycles sold during Bike World's sale week. How many were sold each day?

DAILY SALES—UNITS					
Bicycle	**M**	**T**	**W**	**TH**	**F**
28″ mountain bike	5	8	6	–0–	9
10-speed	11	10	14	12	15
12-speed	15	16	18	17	24
20″ BMX	4	–0–	3	3	5
12″ RMX	2	2	0	1	7
Total					

2. Complete the weekly production report. How many units were produced in one week by each employee?

WEEKLY PRODUCTION REPORT						
Employee Number	**Number of Units**					**Total Units**
	M	**T**	**W**	**TH**	**F**	
31045	65	62	67	61	64	___
10641	63	58	60	61	59	___
42352	54	53	60	59	58	___
10914	68	74	70	76	71	___
91837	59	58	60	60	61	___

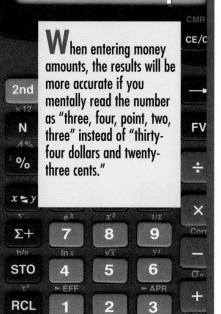

3. Determine how much was spent each month for utilities. Double-check your addition by adding in the reverse order. Note: You must align the decimal points and include a dollar sign in your answer.

UTILITY BILLS

Item	January	February	March	April	May	June
Gas	$ 92.53	$102.10	$ 73.90	$ 54.35	$ 44.11	$ 43.08
Electric	78.14	77.56	74.51	93.04	113.25	137.34
Water	–0–	58.97	–0–	–0–	87.43	–0–
Telephone	26.80	55.20	62.15	42.32	74.10	51.13
Total	_____	_____	_____	_____	_____	_____

4. Complete the production report. Check your totals by adding in the reverse order.

UNIT PRODUCTION REPORT

Name	July 1–7	July 8–14	July 15–21	July 22–28
Sanchez	760	1,212	836	975
Mlynek	685	790	1,005	930
Simpson	725	935	824	1,207
Sumpter	760	1,230	980	874
Woods	770	1,120	915	1,245
Total	_____	_____	_____	_____

5. What was the total amount of the payroll for each of the weeks shown? Remember to align the decimals.

NET PAYROLL REPORT

Name	July 1–7	July 8–14	July 15–21	July 22–28
Sanchez	$280.20	$446.84	$308.22	$359.46
Mlynek	252.55	291.26	370.52	342.87
Simpson	267.29	344.72	303.79	445.00
Sumpter	280.20	453.48	361.31	322.23
Woods	283.88	412.92	337.34	459.01
Total	_____	_____	_____	_____

6. Find the total sales for each day shown on the sales report. Check your answers by adding in the reverse order.

WEEKLY SALES REPORT

Item	May 23	May 24	May 25	May 26	May 27
Decals	$58.00	$61.00	$43.00	$49.00	$36.00
Pennants	63.00	70.00	54.25	57.75	49.00
T-shirts	69.00	57.00	63.00	42.00	39.00
Coffee mugs	55.50	45.00	37.50	42.00	15.00
Steins	41.25	46.75	30.25	33.00	33.00
Total	_____	_____	_____	_____	_____

Solve. Check your answers by adding in the reverse order.

7. Craig Hughes purchased $189 worth of supplies on Monday, $89 on Thursday, and $208 on Friday. Find the total amount of supplies Craig purchased.

8. Gifts & Stuff's sales for the last quarter of the year were as follows: October, $28,774; November, $60,899; December $108,448. What were the total sales for the last quarter?

9. The amount of cash at the end of the day in each cash register of Ben's Grocery is $562, $398, $190 and the amount of cash in the vault is $49,500. What is the total amount of cash on hand?

Activities 1.5

Graph Activity

What is the total amount that Steven Spielberg's top ten films grossed? Write this amount in both digits and words.

_____

Spielberg treasury
Steven Spielberg's *Saving Private Ryan* has grossed over $125 million[1] since its U.S. opening July 24. His top 10 films in worldwide gross (in millions):

Jurassic Park (1993)	$914
E.T. (1982)	$730
The Lost World: Jurassic Park (1997)	$614
Jaws (1975)	$471
Indiana Jones and the Last Crusade (1989)	$450
Raiders of the Lost Ark (1981)	$384
Indiana Jones and the Temple of Doom (1984)	$333
Schindler's List (1993)	$317
Close Encounters of the Third Kind (1977)	$300
Hook (1991)	$277

1 – Through 8/16/98

Source: Exhibitor Relations   By Anne R. Carey and Genevieve Lynn, USA TODAY

Challenge Activity

A business advertises a family recreational boat for lake use. The advertisement states that, if punctured, the boat can get the family to shore before it sinks. The boat can hold 270 liters of water before sinking. It will gain 10 liters of water the first mile, 15 liters the second mile, 20 liters the third mile, and so on. How far can the boat travel before sinking?

Internet Activity

Pretend that this summer's vacation requires getting some specialized biking gear. The products you might need are at **http://www.aardvarkcycles.com/** Click on **site search** and type **avid rollamajig** as your keyword. Just jot down what you want and then add up the order. This company adds a base shipping charge of $4.00 for orders 0–$40. If your order exceeds $40, add an additional $2.00.

SKILLBUILDER 1.6
Adding Horizontally and Vertically

Learning Objectives

* Add whole numbers horizontally.
* Check addition by adding horizontally and vertically.

Horizontal Addition

Figures written in a horizontal line rather than in a column can be rewritten in a vertical column before adding, but this is not always necessary.

Horizontal addition is performed in the same way as vertical addition. Each digit is added to its counterpart in each number: units are added to units, tens to tens, hundreds to hundreds, and so on.

EXAMPLE

Add horizontally: 157 + 316 + 942.

SOLUTION

First add the units digits. Add the *1* (the number carried) to the tens digits. Add the *1* (the number carried) to the hundreds digits.

$$157 + 316 + 942 = 1{,}415$$

$$7 + 6 + 2 = 15 \text{ units}$$

$$1 + 5 + 1 + 4 = 11 \text{ tens}$$

$$1 + 1 + 3 + 9 = 14 \text{ hundreds}$$

14 hundreds + 1 ten + 5 units = 1,415

Self-Check
Add the numbers from left to right. Check by adding from right to left.

23 + 56 + 37 + 81 + 19

MATH TIP

When adding horizontally, be careful to match the proper digits. Horizontal addition can be checked by adding the numbers in the reverse order.

Self-Check Answer
216

Problems

Add the numbers in these problems from left to right. Check your totals by adding from right to left.

1. 6 + 3 + 4 + 9 + 2

2. 57 + 19 + 82 + 36 + 96

3. 5 + 8 + 1 + 7 + 5

4. 54 + 22 + 10 + 83 + 44

5. 7 + 4 + 6 + 3 + 9

6. 99 + 14 + 35 + 23 + 36

7. 8 + 4 + 5 + 6 + 2

8. 139 + 643 + 435 + 815

Horizontal and Vertical Addition

Many business forms require numbers to be added both horizontally and vertically. For example, the following problems are excerpts from ledgers. The total of all the vertical columns should equal the total of all the horizontal rows. This procedure, called **crossfooting,** is the most positive check for accuracy.

EXAMPLE

Add the numbers horizontally and vertically. Then add the totals of the vertical columns and horizontal rows.

234	198	3,024	
2,519	335	488	
307	882	1,360	

SOLUTION

234	198	3,024	3,456
2,519	335	488	3,342
307	882	1,360	2,549
3,060	1,415	4,872	9,347

5	4	3	9	21
7	8	1	2	18
4	7	9	3	23
5	1	6	8	20
21	20	19	22	82

Problems

Add the numbers horizontally and vertically. Then add the totals of the vertical columns and horizontal rows.

9.

16	89	57	
45	65	55	
93	22	18	
44	39	56	

10.

1,526	142	901	
803	345	541	
2,000	93	488	
826	329	281	

11.

159	345	80	
1,345	876	1,250	
947	3,222	435	
500	1,303	612	

Business Applications 1.6

1. On the following sales report, find: the daily total of all the sales; the total for the week for each sales representative; and the grand total for the week.

WEEKLY SALES REPORT

Sales Representative	SU	M	T	W	TH	F	SA	TOTAL
Ling	$2,409	$ –0–	$2,712	$ 3,930	$ –0–	$ 2,664	$ 2,670	_____
Wolf	–0–	3,750	–0–	2,625	2,790	2,889	3,450	_____
Bookner	2,727	3,690	–0–	–0–	3,450	–0–	2,535	_____
Ameredes	2,295	–0–	3,855	2,616	–0–	2,958	2,796	_____
Perez	–0–	2,631	–0–	2,895	2,205	3,600	2,682	_____
Total	_____	_____	_____	_____	_____	_____	_____	_____

Name: _____ Date: _____

2. Find the total number of hours worked each month. Then find the 6-month total for each worker, as well as the total hours worked by all the workers in the 6-month period.

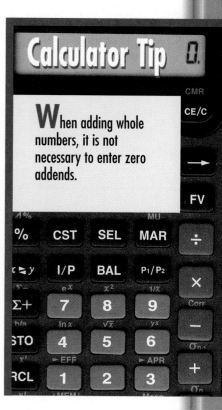

Calculator Tip

When adding whole numbers, it is not necessary to enter zero addends.

HOURS WORKED

Worker	January	February	March	April	May	June	Total
Arnold	168	152	176	168	168	176	____
Bernard	173	160	176	168	168	176	____
Einhorn	168	160	176	173	168	186	____
Grand	168	165	181	168	168	176	____
Johnson	168	165	186	178	168	181	____
Singer	168	160	168	164	168	176	____
Victor	168	160	186	168	168	181	____
Total	____	____	____	____	____	____	____

3. Complete the sales report. Check your totals by adding in the reverse order.

SALES REPORT

Dept.	Cash Sales	Charge Sales	Totals
Suits	$18,565	$ 1,345	_____
Sport Coats	26,673	14,680	_____
Overcoats	9,487	2,450	_____
Trousers	15,432	13,659	_____
Shoes	8,917	8,167	_____
Accessories	783	562	_____
Total	_____	_____	_____

Activities 1.6

Graph Activity

Without using a calculator, compute the total number of vetoes of all the presidents below. _____

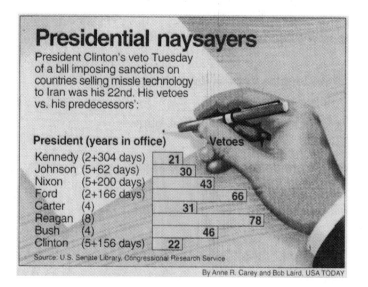

Presidential naysayers

President Clinton's veto Tuesday of a bill imposing sanctions on countries selling missle technology to Iran was his 22nd. His vetoes vs. his predecessors':

President (years in office)	Vetoes
Kennedy (2+304 days)	21
Johnson (5+62 days)	30
Nixon (5+200 days)	43
Ford (2+166 days)	66
Carter (4)	31
Reagan (8)	78
Bush (4)	46
Clinton (5+156 days)	22

Source: U.S. Senate Library, Congressional Research Service

By Anne R. Carey and Bob Laird, USA TODAY

Challenge Activity

Make a chart of all the activities you perform in a week (Monday through Sunday). (You may want to record the number of hours you are in class and the number of hours you spend studying each day.) Then total the hours spent performing each activity all week. Then total the number of hours spent performing all activities each day. Find the total number of hours spent performing all activities in one week.

Internet Activity

Namoeata soaps are handcrafted in the Hawaiian islands. If you were going to make an order of Starburst Soaps, how much would it cost to get a round Morning Star, an oval Morning Star, a Gardenia, a round Makani, and an oval Makani?

http://ssl.maui.net/~namoeata/ order.html

Name:_____ Date:_____

SKILLBUILDER 1.7

Increasing Speed in Adding Large Groups of Numbers

Learning Objectives

* Use subtotals to increase speed in addition.
* Use the accountant's method of addition to add large groups of numbers.

Use Smaller Groups of Numbers

One way to add large groups of numbers is to divide the numbers into smaller groups. That is, break a large group of numbers into two or more subgroups and add each of these subgroups. Then add the totals of the subgroups to find the grand total.

EXAMPLE

Find the total of 428, 645, 27, 156, 42, 583, 897, 273, 6, and 913.

SOLUTION

Write the numbers in a column and break them into two subgroups.

```
428
645
 27
156
 42      a. ___1,298___

583
897
273
  6
913      b. ___2,672___

Total    c. ___3,970___
```

Self-Check
Use subgroups to add: 563 + 294 + 18 + 761 + 109 + 386 + 74 + 3,602 + 866 + 413.

Self-Check Answer
7,086

Problems

Add the numbers in these problems by finding subtotals at the places indicated and then by finding the overall total.

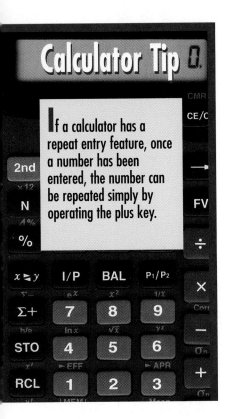

Calculator Tip

If a calculator has a repeat entry feature, once a number has been entered, the number can be repeated simply by operating the plus key.

1. 1,687
    765
     45
  4,525
  <u>8,441</u>
    893
    757
  5,520
     43
  <u>3,192</u>

2. 3,786
  4,508
  1,209
  7,459
  <u>6,663</u>
  2,092
  5,698
  4,983
  3,678
  <u>1,954</u>

3. 15,672
  8,534
  1,345
  87,354
  <u>30,496</u>
  4,598
  15,927
  48,453
  4,899
  <u>41,746</u>

4. 609,576
  198,075
  1,246,934
  4,508
  <u>246,055</u>
  42,233
  120,054
  438
  46,196
  <u>808,417</u>

Answers

1. a. _____

   b. _____

   c. _____

2. a. _____

   b. _____

   c. _____

3. a. _____

   b. _____

   c. _____

4. a. _____

   b. _____

   c. _____

5.   555,050
   1,246,583
      52,176
     459,021
     177,749
   2,243,265
     568,000
       3,491
      44,400
     487,237

Answers

5. a. _____

b. _____

c. _____

Use the Accountant's Method

Another way to break down large groups of numbers is the **accountant's method of addition.** Each column of digits is added separately, and then the column totals are added.

EXAMPLE

Use the accountant's method to add: 4,629 + 816 + 3,937 + 1,014.

SOLUTION

Add the units first (9 + 6 + 7 + 4 = 26), the tens next (2 + 1 + 3 + 1 = 7), the hundreds next (6 + 8 + 9 + 0 = 23), and then the thousands (4 + 3 + 1 = 8). Finally, add the column totals.

$$
\begin{array}{r}
4,629 \\
816 \\
3,937 \\
+1,014 \\
26 \\
07 \\
2\ 3 \\
8 \\
\hline
10,396
\end{array}
$$

Note that the column totals are staggered. When a column total is less than 10, put a zero in front of the number to ensure that the columns are aligned properly. This method of addition is particularly useful in work that is frequently interrupted.

Self-Check
Use the accountant's method to add:
503 + 738 + 4,216 + 59 + 870 + 1,518.

Self-Check Answer
7,904

Problems

Add the numbers in these problems using the accountant's method of addition.

6.
```
   657
 2,097
    43
 4,045
   293
 1,156
```

7.
```
 5,090
 1,534
    54
   441
 2,000
 4,080
```

8.
```
10,921
 5,032
   545
22,318
 4,112
 1,401
```

9.
```
   602
34,839
 3,034
 6,721
    19
 3,501
```

Answers

6. _____

7. _____

8. _____

9. _____

Name:_____ Date:_____

Business Applications 1.7

1. On the following weekly report, find:
   a. The weekly total for each department for each shift.
   b. The day total of all sales for the day for each shift.
   c. The weekly total for each shift.
   d. The day total of all sales for the week.
   Note: The total of the Day Total row should equal the total of the Weekly Total column.

WEEKLY SALES REPORT

Department	SU	M	T	W	TH	F	SA	Weekly Total
Clothing	$ 453	$ 395	$ 404	$ 328	$ 420	$ 450	$ 615	_____
Pharmacy	320	374	291	298	312	345	380	_____
Hardware	95	114	105	83	118	154	165	_____
Notions	192	183	156	133	94	190	125	_____
Sports	108	95	82	211	106	174	115	_____
Shift 1 Total	_____	_____	_____	_____	_____	_____	_____	_____
Clothing	729	1,105	865	1,346	1,428	1,515	1,862	_____
Pharmacy	540	495	611	502	1,003	908	1,432	_____
Hardware	1,230	765	840	903	745	1,762	2,145	_____
Notions	659	583	490	553	429	593	805	_____
Sports	730	1,125	980	1,358	949	1,458	1,652	_____
Shift 2 Total	_____	_____	_____	_____	_____	_____	_____	_____
Clothing	564	435	446	358	475	497	712	_____
Pharmacy	482	654	593	605	983	853	727	_____
Hardware	1,154	839	1,157	720	835	1,156	1,341	_____
Notions	634	602	545	524	487	627	790	_____
Sports	874	972	1,382	892	869	821	945	_____
Shift 3 Total	_____	_____	_____	_____	_____	_____	_____	_____
Day Total	_____	_____	_____	_____	_____	_____	_____	_____

Activities 1.7

Graph Activity

What was the total number of people served (in millions) at the six top libraries? _____

Where books fly off shelves

Patrons checked out more books at the Queens, N.Y., library in 1997 than any other public library. Top libraries in circulation, with population served:

Library (people served, in millions)	Books (in millions)
Queens Borough (1,950,000)	15.3
Los Angeles County (3,324,000)	14.2
Cincinnati/Hamilton County (866,228)	12.6
King County (Seattle) (1,012,620)	12.6
Columbus, Ohio (743,640)	11.9
New York City (3,070,302)	11.1

Source: Public Library Association 1997 Statistical Report

By Anne R. Carey and Genevieve Lynn, USA TODAY

Challenge Activity

Find four members of your class to form your team. Each member will keep track of his or her total expenses for one week. Expenses will include food, housing, medical, clothing, transportation, and childcare. At the end of the week, calculate what the total team expenses were. First calculate each category, then calculate the total expenses.

Internet Activity

County businesses, by type of business, are listed by the census bureau at **http://www.census.gov/ pub/epcd/cbp/view/us95.txt**.

The top nine businesses listed employed how many people?

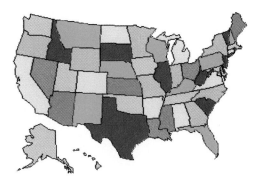

SKILLBUILDER 1.8

Rounding Numbers and Estimating Answers

Learning Objectives

* **Round whole numbers to a specified place.**
* **Estimate a sum using rounded numbers**

Rounding Numbers

Rounding is replacing an actual figure or amount with an approximate figure. For example, sales of 19,883 items could be rounded to sales of around 20,000 items for a verbal report. Money amounts are often rounded to the nearest cent. A sales tax of $0.686 is rounded to $0.69.

To round a number, first locate the place to which the number will be rounded. Then look at the digit to the right of this place. If this digit is 5 or greater, increase the digit in the rounding place by 1. If this digit is less than 5, the digit in the rounding place remains the same. Digits to the right of the rounding place are replaced by zeros.

EXAMPLE

Round 5,627 to the tens place.

SOLUTION

Put a line under the digit in the tens place. Then place an arrow over the digit to the right.

$$\downarrow$$
$$5,6\underline{2}7$$

Since 7 is greater than 5, increase by 1 the digit in the tens place. The tens digit becomes 3 and the units digit is replaced with a zero.

5,627 rounds to 5,630

Self-Check
Round 5,627 to the hundreds place.

MATH TIP

Estimating will let you know if your answer is reasonable. If you want to know if your estimate is correct, check by adding the numbers in a different order.

**Self-Check
Answer**
5,600

Problems

Round each number as indicated.

1. 553 to the nearest ten
2. 8,605 to the nearest hundred
3. 32,609 to the nearest ten
4. 52,999 to the nearest thousand
5. 17,839 to the nearest ten
6. 37,056 to the nearest thousand
7. 2,392 to the nearest hundred
8. 5,949 to the nearest ten
9. 763 to the nearest hundred
10. 547 to the nearest thousand
11. 16,572 to the nearest hundred
12. 46,219 to the nearest ten thousand
13. 5,449 to the nearest ten
14. 237,219 to the nearest thousand
15. 776 to the nearest ten
16. 24,986 to the nearest hundred
17. 17,605 to the nearest ten thousand
18. 4,762 to the nearest thousand

Answers

1. _____
2. _____
3. _____
4. _____
5. _____
6. _____
7. _____
8. _____
9. _____
10. _____
11. _____
12. _____
13. _____
14. _____
15. _____
16. _____
17. _____
18. _____

Estimating Answers in Addition

Estimating answers helps to uncover obvious errors in computations. Being able to estimate answers is especially important when using calculators.

One way to estimate an answer to an addition problem is to round each number to its largest place. Then add the rounded numbers. The estimate will be close to the actual answer.

EXAMPLE

Estimate the sum: 322 + 576 + 349 + 105. Then determine the actual sum.

SOLUTION

Estimate	Actual
300	322
600	576
300	349
100	105
1,300	1,352

Self-Check
Estimate the sum: 2,529 + 1,432 + 770 + 4,200. Then determine the actual sum.

Self-Check Answer
Estimate 8,800
Actual 8,931

Problems

Estimate the answer to each problem by rounding each number to its largest place and adding the estimates. Then add the numbers to determine the actual totals. Compare your estimated answer with your actual answer to see if your actual answer is reasonable.

19. Actual	Estimate	20. Actual	Estimate	21. Actual	Estimate
37		88		47	
61		8		62	
43		52		19	
89		96		39	
52		12		65	
46		68		22	
___		___		___	

22. Actual	Estimate	23. Actual	Estimate	24. Actual	Estimate
737		209		785	
242		454		362	
495		298		411	
315		549		530	
389		897		647	
669		242		645	
___		___		___	

Calculator Tip

Estimate addition before you enter numbers into the calculator to assist in recognizing errors.

25. Actual	Estimate	26. Actual	Estimate	27. Actual	Estimate
3,067		4,381		7,428	
4,198		7,007		3,664	
4,751		3,436		1,283	
2,009		3,605		4,245	
4,113		1,651		6,495	
4,851		1,770		2,629	
___		___		___	

Activities 1.8

Graph Activity

Round New York's per capita personal income to the nearest thousand.

_____

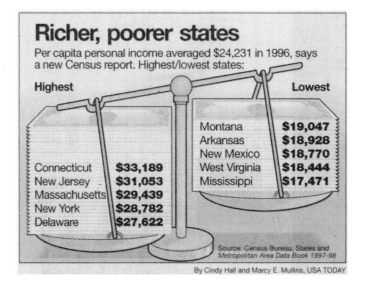

Richer, poorer states

Per capita personal income averaged $24,231 in 1996, says a new Census report. Highest/lowest states:

Highest		Lowest	
Connecticut	$33,189	Montana	$19,047
New Jersey	$31,053	Arkansas	$18,928
Massachusetts	$29,439	New Mexico	$18,770
New York	$28,782	West Virginia	$18,444
Delaware	$27,622	Mississippi	$17,471

Source: Census Bureau, States and Metropolitan Area Data Book 1997-98

By Cindy Hall and Marcy E. Mullins, USA TODAY

Challenge Activity

Save the cash register receipts you receive when buying groceries for one month. How much do you spend on groceries in one month's time? In a week? To determine how much is spent on food, also keep track of the amount spent on eating out. Combine the two amounts to get the total spent.

Internet Activity

How do numbers with decimals get rounded? Try some at

http://www.txdirect.net/users/everett/rounding.html

SKILLBUILDER 1.9

Subtracting Whole Numbers

Learning Objectives

- **Find the difference between two numbers.**
- **Check subtraction by using addition.**

Subtraction

In subtraction, each digit in the **subtrahend** is subtracted from the digit in the same place (that is, having the same place value) in the **minuend.** The result of the subtraction is called the **difference.** If a digit in the subtrahend is larger than the digit above it, it is necessary to borrow before subtracting.

EXAMPLE

Subtract: 74 − 56.

SOLUTION

Ten units are borrowed from the tens column and added to the units column.

```
   6 14
  7̶4̶  Minuend
− 56  Subtrahend
  18  Difference
```

MATH TIP

In order to subtract 56 from 74, you must rename 74 as 7 tens + 4 units = 6 tens + 1 ten + 4 units = 6 tens + 14 units.

Self-Check
Subtract: 325 − 158.

Self-Check Answer
167

Problems

Find the difference in each of these problems.

	a.	b.	c.	d.	e.	f.	g.
1.	97	42	40	61	73	89	65
	44	28	27	39	59	24	25
2.	561	402	852	820	818	211	604
	429	128	749	427	239	106	355

Checking Subtraction

Check subtraction by adding the subtrahend to the difference; the sum should equal the minuend.

EXAMPLE

Subtract 4,580 from 7,870. Check your answer.

SOLUTION

Check

$$
\begin{array}{r}
{}^{7\,17}\\
7,870 \\
-4,580 \\
\hline
3,290
\end{array}
\quad
\begin{array}{l}
\text{Minuend} \\
\text{Subtrahend} \\
\text{Difference}
\end{array}
\qquad
\begin{array}{r}
4,580 \\
+3,290 \\
\hline
7,870
\end{array}
\quad
\begin{array}{l}
\text{Subtrahend} \\
\text{Difference} \\
\text{Minuend}
\end{array}
$$

Self-Check
Subtract: 1,398 − 799. Check your answer.

Self-Check Answer 599

© Glencoe/McGraw-Hill

Problems

Find the difference in each problem. Check by adding the subtrahend to the difference.

	a.	b.	c.	d.	e.
3.	1,807 643	436 278	9,002 4,928	954 643	6,598 2,693
4.	15,087 13,258	47,931 28,853	356,803 58,999	134,628 99,999	67,550 32,536
5.	3,907 −948	810 −280	9,407 −1,712	7,236 −289	5,933 −3,721
6.	8,079 −555	423 −294	679 −194	25,003 −2,735	86,004 −9,627
7.	8,202 −3,917	7,206 −1,624	43,748 −2,367	8,237 −62	27,034 −8,237
8.	80,529 −21,076	12,701 −9,514	3,933 −387	95,442 −86,857	6,465 −2,975
9.	3,050,092 −1,552,325		8,000,235 −1,230,768	7,826,532 −3,491,673	2,425,324 −342,259

Business Applications 1.9

1. **Net sales** is the difference between sales and sales returns. Determine the net sales for each day of the week for Essex Street Market.

ESSEX STREET MARKET

	SU	M	T	W	TH	F
Sales	$10,943	$8,678	$7,586	$8,840	$9,624	$12,955
Sales returns	217	642	258	225	195	381
Net sales	_____	_____	_____	_____	_____	_____

Solve.

2. How much more are the cash receipts of $3,236 on July 2 than the $1,872 receipts on July 4?

3. John bought a CD ROM encyclopedia for $132 and received a $39 rebate. How much did John pay for the encyclopedia?

4. Fred is planning a trip to Lake Placid. The distance between Jacksonville, Florida and Lake Placid, New York is 1,230 miles. If Fred drove 689 miles from Jacksonville toward Lake Placid on the first day, how many more miles are left to drive?

Answers

2. _____

3. _____

4. _____

Activities 1.9

Graph Activity

What is the difference in the amount spent on ads in health care publications and business publications? _____

Trade publication ads up

Ad spending in business-to-business publications rose 12% to $8 billion in 1997. Leading ad categories:

Technology
$3.5 billion

Health care/drugs
$920 million

Business/financial
$474 million

Manufacturing/processing/industrial
$434 million

Business travel
$362 million

Source: Business Information Network (BIN)   By Anne R. Carey and Bob Laird, USA TODAY

Challenge Activity

The administrative offices of Toko Computer Company had a petty cash fund with a beginning balance of $350. Withdrawals from the fund were $35, $92, $15, $120, $12, $9, and $41 for the month, and deposits were $150. What was the balance in the petty cash fund at the end of the month?

Internet Activity

Try your hand at the ancient abacus method of subtraction at the interactive site,

http://www.ee.ryerson.ca/~elf/ abacus/asubtraction.html

Upper deck — Beads
— Rods

Lower deck — Frame

SKILLBUILDER 1.10

Subtracting Horizontally and Estimating Answers

Learning Objectives

- Subtract vertically and horizontally.
- Estimate answers in subtracting using rounded numbers.

Subtracting Horizontally

In horizontal subtraction, each digit in the subtrahend is subtracted from the digit in the same place in the minuend, units from units, tens from tens, and so on.

   Subtraction can also be checked by computing both horizontally and vertically. The total of the horizontal minuends *less* the total of the horizontal subtrahends should equal the total of the horizontal differences.

EXAMPLE

Subtract horizontally. Then find vertical column totals and subtract.

$$15 - 5$$
$$20 - 3$$
$$56 - 4$$

SOLUTION

$$15 - 5 = 10 \qquad 15 - 5 = 10$$
$$20 - 3 = 17 \qquad 20 - 3 = 17$$
$$56 - 4 = 52 \qquad \underline{56} - \underline{4} = \underline{52}$$
$$ 91 - 12 = 79$$

MATH TIP

The decimal point is understood in a whole number; therefore, the decimal point does not have to be written. We must, however, remember the decimal point is to the right of the number when solving problems.

Problems

Complete the following records and forms by subtracting horizontally.

1.

BIKE WORLD WEEK OF APRIL 23, 20—			
Item	**Beginning**	**End**	**Sales**
28″ mountain bike	40	12	__
10-speed	90	28	__
12-speed	125	35	__
20″ BMX	45	30	__
12″ RMX	35	23	__

2.

CHILDREN'S WEAR MONTH OF AUGUST, 20—			
Item	**Beginning**	**End**	**Sales**
Blouses	398	98	___
Dresses	435	124	___
Jackets	145	53	___
Jeans	503	218	___
Shirts	258	58	___
Shorts, Boys'	382	151	___
Shorts, Girls'	312	144	___
Skirts	482	224	___
Socks	505	132	___
Sweaters	165	76	___

3.

ACCOUNTS RECEIVABLE REPORT SEPTEMBER 30, 20—			
Account Number	**Billed**	**Discount Allowed**	**Received**
115	$467.98	$18.72	_____
125	190.23	3.80	_____
155	102.45	3.07	_____
165	93.25	0.93	_____
185	303.46	9.10	_____
195	88.72	0.89	_____

4.

ACCOUNTS PAYABLE REPORT SEPTEMBER 30, 20—			
Account Number	**Amount Due**	**Amount Paid**	**Balance Due**
230	$1,456	$ 956	____
240	890	250	____
250	2,312	1,312	____
260	563	250	____
270	934	450	____
280	450	225	____

Name:_____ Date:_____

5. Ernest Enterprises maintains a check register in which the company records all checks that it writes to pay for invoices for the merchandise it has purchased. The amount of the check is the amount of the invoice less any cash discount. Complete the check register and check your computations. Note: you must align the decimals.

CHECK REGISTER
ERNEST ENTERPRISES
JUNE 17, 20—

Check Number	Amount of Invoice	Cash Discount	Amount of Check
577	$ 357.63	$ 10.73	_____
578	1,982.75	99.14	_____
579	145.90	2.92	_____
580	55.67	–0–	_____
581	598.50	29.93	_____
582	346.38	10.39	_____
583	28.25	–0–	_____
584	481.09	19.24	_____
Total	_____	_____	_____

6. Complete the inventory and check your computations.

BIKE WORLD
INVENTORY FOR WEEK ENDING APRIL 23, 20—

Item	Value of Beginning Inventory	Amount of Sales	Value of Inventory on Hand
28″ mountain bike	$ 3,399.60	$2,379.72	_____
10-speed	8,099.10	5,579.38	_____
12-speed	12,811.25	9,224.10	_____
20″ BMX	3,143.25	1,047.75	_____
12″ RMX	1,748.95	599.64	_____
Total	_____	_____	_____

7. Complete the inventory and check your computations. Remember to align decimals.

SANDY'S SPORTSWEAR INC.
INVENTORY FOR WEEK ENDING JULY 11, 20—

Item	Value of Beginning Inventory	Amount of Sales	Value of Inventory on Hand
Blouses	$2,358.09	$954.65	_____
Tank Tops	987.87	254.98	_____
Jeans	3,459.22	1,040.75	_____
Shoes, Canvas	1,009.45	550.34	_____
Shoes, Leather	2,104.50	973.62	_____
Shorts	1,985.02	654.82	_____
Skirts	3,987.38	1,269.02	_____
Tennis dresses	1,753.72	703.45	_____
Total	_____	_____	_____

8. The following charts show the number of business starts and business failures for the months Jan. to June and July to Dec. in three regional areas. For each region, determine the difference in business starts and failures.

Business Starts	Jan. to June	July to Dec.	Difference
Region A	4,964	5,455	_____
Region B	19,877	21,708	_____
Region C	39,966	41,948	_____
Total	_____	_____	_____

Business Failures	Jan. to June	July to Dec.	Difference
Region A	463	357	_____
Region B	1,733	1,514	_____
Region C	4,070	3,998	_____
Total	_____	_____	_____

Estimating Answers in Subtraction

Answers to subtraction problems can be estimated by rounding each number to its largest place and then subtracting the rounded numbers. The estimate will be close to the actual answer.

EXAMPLE

Estimate the difference: 29,689 − 12,271. Then find the actual answer.

SOLUTION

Estimate	Actual
30,000	29,689
−10,000	−12,271
20,000	17,418

Self-Check
Estimate the difference: 39,872 − 6,819. Then find the actual answer.

MATH TIP

Remember, an estimate will not let you check the accuracy of specific digits in your answer, but it will give you an idea of whether your exact answer is reasonable.

Self-Check
Answer
Estimate 33,000
Difference 33,053

Problems

Estimate the answer to each problem below by rounding each number to its largest place and then subtracting. Then find the actual difference.

9.		10.		11.	
Actual	*Estimate*	*Actual*	*Estimate*	*Actual*	*Estimate*
38,796		36,329		26,392	
−25,432		−5,124		−9,684	

12.		13.		14.	
Actual	*Estimate*	*Actual*	*Estimate*	*Actual*	*Estimate*
73,256		41,630		11,651	
−38,197	_____	−7,814	_____	−3,597	_____

Calculator Tip

Addition and subtraction can be done within the same problem. Subtract by pressing the minus sign.

Activities 1.10

Graph Activity

What is the difference in the amount of the average monthly cost of day care in Boston and Manchester? _____

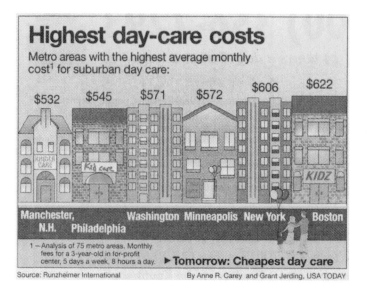

Highest day-care costs

Metro areas with the highest average monthly cost[1] for suburban day care:

$532 $545 $571 $572 $606 $622

Manchester, N.H.   Washington   Minneapolis   New York   Boston
Philadelphia

1 – Analysis of 75 metro areas. Monthly fees for a 3-year-old in for-profit center, 5 days a week, 8 hours a day.   ▶ Tomorrow: Cheapest day care

Source: Runzheimer International   By Anne R. Carey and Grant Jerding, USA TODAY

Challenge Activity

Use newspaper supermarket ads as a guide in making a shopping list. Choose items that appear in more than one supermarket's ad. Write the advertised price for 2 to 3 supermarkets. Determine if one store is consistently less expensive. Decide where to shop this week to save the most money.

Internet Activity

A little subtraction (and some other math operations) go a long way in this brainteaser:

http://pages.cthome.net/whiplash/trick.txt

SKILLBUILDER 1.11

Problem Solving

Learning Objectives

* Solve a word problem using problem-solving steps.
* Solve business applications involving addition and subtraction.

Using Problem-Solving Steps

An important factor in solving word problems is knowing when to add, subtract, multiply, or divide. Often key words in the problem will indicate which operations are required. For example, *of, total,* and *how many in all* often indicate multiplication. *For each, per,* and *find the average* usually indicate division. *Plus, total, sum,* and *how many in all* often indicate addition. *Amount of increase, amount of decrease, less, difference, balance,* and *how many more* are words that often indicate subtraction. The word *what* indicates the unknown factor; *is* indicates equals. Examining these key words will help in expressing a word problem as a number problem.

To solve a word problem, read the problem carefully, more than once if necessary. Then follow these steps.

1. Determine what the problem is asking you to find.

2. Decide what information is necessary in order to solve the problem.

3. Decide what arithmetic operation to use (look for key words).

4. Write a problem and solve it.

5. **Reread** the problem to make sure that your answer is sensible. Then check the answer.

> **MATH TIP**
>
> When solving a word problem, be sure your answer is accurate. That is, check your work using the methods of checking you have learned. Then be sure your answer is sensible—for instance, a unit cost of $10 for a bicycle is not sensible and indicates an error in the solution process.

EXAMPLE

Elbert Products spent $1,800 on a holiday party for their employees, $2,100 on gifts for clients, $18,000 on employee bonuses, and $346 on holiday decorations. How much did they spend in all?

SOLUTION

1. What is the problem asking you to find?
   Total holiday expenses

2. What is the necessary information?
   Party: $1,800   Gifts: $2,100
   Bonuses: $18,000   Decorations: $346

3. Key Words: *in all*   (indicates addition)

4. Write a problem and solve it:
   $1,800
   2,100
   18,000
   346
   ―――
   $22,246

5. Reread the problem and check by adding in reverse order.
   Elbert Products spent $22,246 in all on holiday expenses.

**Self-Check
Answer**
304 kg

Self-Check
Five crates of parts are to be shipped by National Airways. The crates weigh 56 kg, 46 kg, 74 kg, 68 kg, and 60 kg. What is the total weight of the crates to be shipped?

© Glencoe/McGraw-Hill

Name:_____ Date:_____

Problems

Solve these problems using the problem-solving steps as a guide.

1. Daniel Durant, the maintenance plumber at the Sterling Apartments, had to replace damaged pipes in the sprinkler system. He cut and installed five pieces of pipe, measuring 8 ft, 3 ft, 2 ft, 9 ft, and 6 ft. How many feet of pipe in all are needed to repair the sprinkler system?

2. In 1997, 12,500,000 passengers used Main Terminal Air facilities. By 2000, the number had risen to 30 million passengers per year. What is the total increase in passenger use from 1997 to 2000?

2. _____

3. A total of 5,828 people toured the plant facilities at South Service Company during their annual open house. This is 378 fewer people than toured during last year's open house. How many people toured last year?

3. _____

4. Fashion-tec has a production capacity of 35,000 units a week. Their actual production last week was Monday, 3,274 units; Tuesday, 2,492 units; Wednesday, 3,194 units; Thursday, 6,436 units; and Friday, 5,432 units. What was the difference between their actual production and their production capacity?

4. _____

Student Success Hints

Use an estimate as well as common sense to determine if an answer is reasonable. Then check your computations.

Problem-Solving in Business

Business transactions are often recorded in a record called an **account.** An account resembles the letter T. The left side of the account is called the **debit** side; the right side of the account is called the **credit** side. To find how much is in the cash account, add all the debit amounts; then add all the credit amounts. The smaller number is subtracted from the larger number. If the total of the debit side is larger, the account has a **debit,** or negative, **balance.** If the total on the credit side is larger, the account has a **credit,** or positive, **balance.**

EXAMPLE

Determine whether there is a debit or credit balance for debits of $160.00, $2,650.50, $65.45, $492.11, and $745.73 and credits of $56.25, $643.90, $85.00, and $234.75.

SOLUTION

Cash

Debit	Credit
160.00	56.25
2,650.50	643.90
65.45	85.00
492.11	234.75
745.73	$1,019.90
4,113.79	

$4,113.79
−1,019.90
$3,093.89   Debit balance

Note that debits are greater than credits, so the balance is a debit balance.

Self-Check Answer

Cash

Debit	Credit
$ 35.88	$1,089.34
1,405.90	42.76
332.19	882.10
38.55	$2,014.20
$1,812.52	

$2,014.20 − $1,812.52 = $201.68 Credit Balance

Self-Check

Determine whether there is a credit or debit balance for debits of $35.88, $1,405.90, $332.19, and $38.55 and credits of $1,089.34, $42.76, and $882.10.

Problems

5. Add the debit and credit amounts in the accounts shown below. Subtract and determine whether the accounts have debit or credit balances.

Accounts Receivable		Supplies		Equipment		Accounts Payable	
Debit	**Credit**	**Debit**	**Credit**	**Debit**	**Credit**	**Debit**	**Credit**
5,678.90	790.00	150.00	50.00	7,000.00	1,000.00	1,025.00	5,200.00
790.00	2,678.90	65.00	6.75	950.00	450.00	340.00	680.00
1,434.08	434.08	93.56	12.80	10,450.25	4,050.00	340.00	575.50
53.25	500.00	29.04		700.00		2,600.00	590.00
2,000.00	53.25	6.85		3,450.50			3,090.75
194.34		42.09		1,268.60			1,258.05
53.25							

6. A cashier should make change in the fewest number of coins or bills possible. Use one dime, not two nickels. Use a $10 bill, not two $5 bills. Find the change for these purchases. Then select the correct coins and bills to make up the change.

	Amount Given In Payment	Amount of Purchase	Amount of Change
Ex.	$10.00	$2.58	$7.42
a.	5.00	1.49	_____
b.	7.00	6.43	_____
c.	20.00	5.04	_____
d.	15.00	12.51	_____
e.	1.00	.59	_____
f.	50.00	29.78	_____

1¢	5¢	10¢	25¢	$1	$5	$10	$20
2	1	1	1	2	1		

Business Applications 1.11

Answers

1. _____

_____

2. _____

3. _____

4. _____

_____

Calculator Tip

When solving word problems involving whole numbers, first estimate the answer. Then set the decimal selector to zero and work the problem on the calculator.

1. Hopkins Medical Equipment and Supplies showed the following sales and returns for two days of business. Monday: sales of $73.19, $25.40, $324.65, and $8.49 and returns of $20.77 and $56.82; Tuesday: sales of $38.89, $42.57, $17.33, and $168.55 and returns of $12.39. Determine the net sales for the two days. Did Hopkins Medical Equipment and Supplies have a debit or credit balance for the two days?

2. Trisha's take home pay is $3,230 per month. After she makes her mortgage payment of $1,190, how much does she have left?

3. Media Play had $14,950 in discontinued inventory. After their discontinued inventory sale, there was $3,590 left. How much did they sell?

4. Marcia Hart bought a cabin on the lake for $53,000. Her father gave her $10,000 and her aunt gave her $5,000 to use as a down payment. She withdrew another $4,000 from her savings. The seller allowed her to pay $10,000 of the price in 5 years as a lump sum payment. Marcia agreed to that and financed the remaining amount with a mortgage company. What was the amount of the loan with the mortgage company and what is the total amount Marcia still owes on the cabin?

Name:_____   Date:_____

Activities 1.11

Graph Activity

How many acres of land burned in 1995 through 1997? _____

How bad is the fire season?

Since Memorial Day, fires in Florida have charred 232,985 acres; 38 of 67 counties still have fires. But the 700,000 acres burned nationwide are far fewer than the 2.4 million acres burned by this time in 1996. Wildland fires, 1995–1998:

	Fires	Acres burned
1998 (through 6/28)	29,885	698,387
1997	66,196	2,856,959
1996	96,363	6,065,998
1995	82,234	1,840,546

Source: National Interagency Fire Center, Boise, Idaho; Unified Area Command

By Anne R. Carey and Elys A. McLean, USA TODAY

Challenge Activity

It is important to carefully budget one's income. Make a list of your monthly expenses. List all expenses. Subtract your total expenses from your monthly income. How much money can you save each month for an emergency fund?

Internet Activity

The most personal meaning of the word subtraction is when money is taken out of your bank account, by you, to purchase items. Today, ATM cards and debit cards are highly popular. What are the advantages and disadvantages of using "plastic" to subtract from your account? It's a plus to know:

http://www.insiderreports.com/ bizrprts/nolo88.htm

UNIT 2

WHOLE NUMBERS
Multiplication and Division

The U.S. government gathers data regularly about various aspects of our lives. Data are collected and analyzed for presentation through numerous publications, including almanacs. The partial chart to the right shows the amount of fuel consumed by different modes of ground transportation. In order to find averages, you need to divide.

In this unit we study multiplication and division of whole numbers and money amounts. The business applications of multiplication and division that we consider include invoices, sales records, and averages.

Type of Vehicle	Fuel Consumption (billions of gallons)*	Avg. Fuel Used per Vehicle (gallons)*	Avg. Miles per Gallon*
Cars	72.4	505	21.00
Buses	0.9	1,436	5.36
Trucks	58.1	1,305	10.62

MATH CONNECTIONS

Bank Manager

In the past decade or so, the banking industry has become quite competitive. In order to attract new customers to open an account, banks often run expensive radio, TV, and newspaper ads that may offer free gifts to new customers.

Jane Polk is a branch manager who is paid a base salary as well as a bonus, depending upon the profits of the branch. The bonus is an incentive to make the branch as profitable as possible. The manager must keep track of revenue and expenses in order to make appropriate decisions that will increase the bank's profits (and her paycheck). Jane's duties often involve working with decimals.

Math Application

A particular bank is offering new customers a free toaster when opening an account. The manager predicts that 100 new accounts will be opened due to the offer. If each new account orders checks, from which the bank makes $8.00 profit, how much profit does the bank manager expect to make from the check orders?

$$100(\$8.00) =$$

The manager was right. The free gift offer resulted in 100 new accounts. If the free gift costs the bank $5.00 each, what is the profit the bank receives after the gifts have been purchased? Use the formula to determine the profit.

$$\$800 - 100(\$5) =$$

Critical Thinking Problem

The bank manager decides to give away coffee mugs, decorated with the bank's logo. Company A agrees to prepare the coffee mugs with the logo for $2.50 each. Company B agrees to prepare the coffee mugs for $1.50 each plus a one time set-up fee of $125.00. Which company do you think the manager should choose? (Hint: Think of the number of new accounts projected.)

Multiplication: Finding the Product

Learning Objectives

- Find the product of two whole numbers.
- Find the product of money and a whole number.

Multiplying Whole Numbers

The number by which you multiply is the **multiplier,** the number being multiplied is the **multiplicand,** and the result of the computation is the **product.**

When the multiplier has two or more digits, the multiplicand is multiplied separately by each digit in the multiplier. These products are then added to find the final product. It is important to align the subproducts properly so that the correct answer will be obtained when the products are added.

EXAMPLE

Find the product: 43 × 52.

SOLUTION

```
   43    Multiplicand
  ×52    Multiplier
   86
  215
 2,236   Product
```

Self-Check
Find the product: 104 × 85.

MATH TIP

When you find the product of 43 and 52, you first multiply 43 × 2 and then 43 × 50. Thus, the partial product in the second row is actually 2,150, although the zero is not usually written.

Self-Check Answer
8,840

Answers

1. _____

2. _____

3. _____

4. _____

5. _____

6. _____

7. _____

8. _____

9. _____

10. _____

11. _____

12. _____

13. _____

14. _____

15. _____

16. _____

Problems

Find the products and any totals.

1. $41 \times 5 =$ _____
   $87 \times 4 =$ _____
   $44 \times 3 =$ _____
   $61 \times 7 =$ _____
   $58 \times 9 =$ _____
   $99 \times 2 =$ _____
   Total = _____

2. $359 \times 6 =$ _____
   $7,412 \times 9 =$ _____
   $432 \times 3 =$ _____
   $1,546 \times 7 =$ _____
   $435 \times 8 =$ _____
   $3,452 \times 4 =$ _____
   Total = _____

3. $\begin{array}{r} 78 \\ \times\ 54 \\ \hline \end{array}$

4. $\begin{array}{r} 856 \\ \times\ 32 \\ \hline \end{array}$

5. $\begin{array}{r} 754 \\ \times\ 392 \\ \hline \end{array}$

6. $\begin{array}{r} 8,065 \\ \times\ 44 \\ \hline \end{array}$

7. $\begin{array}{r} 6,543 \\ \times\ 798 \\ \hline \end{array}$

8. $\begin{array}{r} 842 \\ \times\ 53 \\ \hline \end{array}$

9. $\begin{array}{r} 421 \\ \times\ 365 \\ \hline \end{array}$

10. $\begin{array}{r} 5,384 \\ \times\ 779 \\ \hline \end{array}$

11. $\begin{array}{r} 6,739 \\ \times\ 504 \\ \hline \end{array}$

12. $\begin{array}{r} 9,126 \\ \times\ 867 \\ \hline \end{array}$

13. $\begin{array}{r} 1,288 \\ \times\ 7 \\ \hline \end{array}$

14. $\begin{array}{r} 3,492 \\ \times\ 5 \\ \hline \end{array}$

15. $\begin{array}{r} 658 \\ \times\ 36 \\ \hline \end{array}$

16. $\begin{array}{r} 16,928 \\ \times\ 1,007 \\ \hline \end{array}$

Student Success Hints

Learning Styles

Take charge of your learning. Let your instructor know when you don't understand something. Seek help until you do understand it.

Multiplying With Money

When multiplying an amount of money given in dollars and cents by a whole number, the answer is also written in terms of dollars and cents. Remember cents is not a whole number but a decimal.

Name:_____ Date:_____

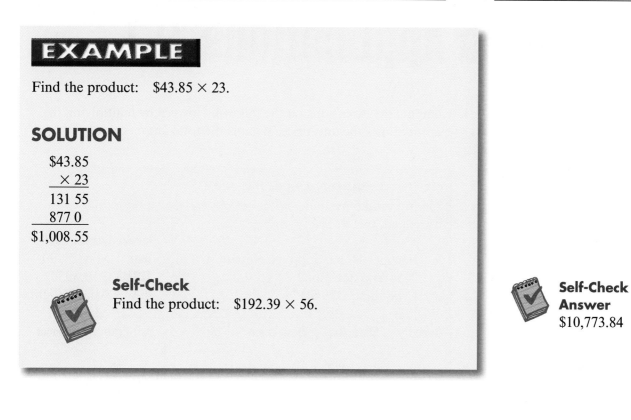

EXAMPLE

Find the product: $43.85 × 23.

SOLUTION

$43.85
  × 23
131 55
 877 0
$1,008.55

Self-Check
Find the product: $192.39 × 56.

Self-Check Answer
$10,773.84

Problems

Find the products.

17.	$23.56	18.	$54.75	19.	$182.09	20.	$265.28
	× 8		× 12		× 28		× 309

Answers

17. _____

18. _____

19. _____

20. _____

Business Applications 2.1

Answers

1. _____

2. _____

1. Compute the extensions on the following invoice by multiplying the quantity times the unit price. What is the total amount due?

GUTFELD HARDWARE COMPANY
234 Mandrake Avenue
Cleveland, OH 44118

Invoice Number 6308

To   Bochner Home Supply Company
2654 Marston Road
Lyndhurst, OH 44124

Date   May 24, 20—
TERMS 2/10, n/30
VIA   Atlas Freight

	Quantity	Stock No.	Description	Unit Price	Amount
a.	24	ED–345	Electric drills—$\frac{3}{8}$ in.	$21.68	_____
b.	54	SS–46S	Socket wrench set—standard	6.50	_____
c.	18	SS–46M	Socket wrench set—metric	6.50	_____
d.	144	LP–10	Locking plier—10 in.	1.50	_____
e.	144	AW–8	Adjustable wrench—8 in.	2.28	_____
				Total	_____

---

2. Bob Dance Honda's Labor Day Sale advertised a dealer inventory reduction sale on 48 Accords. Each car would be marked down to $15,500 and there would be no negotiations or trade-ins. If Bob Dance Honda sold 41 Accords at the advertised price, how much revenue did they make from the sale?

3. Krazee Karl's Music Shop purchased 2,116 tapes and compact discs from a bankrupt music store for $2.25 each. Karl sold them in his own shop at $8.57 each for the 312 tapes. The rest, which were compact discs, were sold for $12.39 each.
   a. How much did Krazee Karl receive from the sale of the 2,116 pieces?

   b. What was the gross profit realized on this sale (total sales − cost of goods = gross profit)?

4. Helena Betz used the Bosco Company car for a business trip of 456 miles. After the trip, the odometer on the company car read 30,524. What was the odometer reading before she left on the trip?

5. Alabiss Kennels boards cats and dogs. The fee for boarding a cat is $5 per day. The fee for boarding a dog is $7 per day for dogs under 20 pounds and $10 per day for dogs 20 or more pounds. On Sunday, there were 8 cats, 12 dogs under 20 pounds, and 3 dogs 20 or more pounds. How much revenue did Alabiss make from boarding cats and dogs on Sunday?

6. Wee Care Day School charges $20 per child for day care and $2 for lunch. On Friday, there were 32 children at the day care and they all had lunch there. How much revenue did Wee Care Day School make on Friday?

Answers

3. a. _____

   b. _____

4. _____

5. _____

6. _____

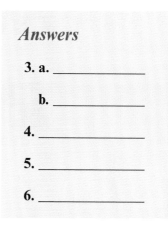

Calculator Tip

To find the product of a number times itself on some calculators, enter the number and press ⊗ and then =.

Activities 2.1

Graph Activity

Look at the graph below. If each ticket in 1998 was sold for $40, how much revenue would be collected? _____

Filling Broadway's seats

The booth in New York's Times Square — TKTS — that offers unsold Broadway and Off Broadway show tickets at half price on the day of performance opened on June 25, 1973. Half-price tickets sold annually[1] since it opened:

1974   450,932
1978   992,975
1983   1,840,581
1988   1,623,982
1993   1,151,325
1998[2]   1,700,000

1 – Estimated total by fiscal year ending June 30.
2 – Estimate

Source: Theatre Development Fund

By Anne R. Carey and Genevieve Lynn, USA TODAY

Challenge Activity

Mickey's T-Shirt Shop is having a T-shirt sidewalk sale. She needs $50 in coins and one-dollar bills for the cashier. What different combinations of pennies, nickels, dimes, quarters, and one-dollar bills would you suggest?

Internet Activity

What is your cardiac output? Take your heart rate and multiply it by milliliters per beat, as suggested at the following website.

http://quest.arc.nasa.gov/neron/video/experiments/get_pumped.html

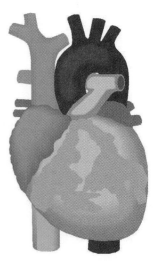

SKILLBUILDER 2.2

Checking and Estimating With Multiplication

Learning Objectives

- **Check the accuracy of multiplication by interchanging factors.**
- **Estimate products using rounded numbers.**

Checking the Accuracy of Multiplication

To check the accuracy of multiplication, interchange the *factors* (the multiplier and the multiplicand) and multiply again. The product should be the same no matter what order the factors are multiplied.

EXAMPLE

Multiply: 597×63. Check by interchanging factors.

> **MATH TIP**
>
> *When multiplying two whole numbers, the product will be greater than the multiplier or multiplicand if neither is 1 or 0.*

SOLUTION

```
    597          Check
  × 63              63
  1 791          × 597
 35 82             441
 37,611          5 67
                 31 5
                 37,611
```

 Self-Check
Multiply: 318×274. Check by interchanging the factors.

 Self-Check Answer
87,132

Problems

Find the product for each of the following problems. Then check your answer by reversing the order of the numbers and multiplying again. Show your work.

1. a.　　　　**Check**　　**b.**　　　　**Check**　　**c.**　　　　**Check**

$$\begin{array}{r} 95 \\ \times\ 23 \\ \hline \end{array}$$ 　　　　$$\begin{array}{r} 358 \\ \times\ 74 \\ \hline \end{array}$$ 　　　　$$\begin{array}{r} 716 \\ \times\ 48 \\ \hline \end{array}$$

2. a.　　　　**Check**　　**b.**　　　　**Check**　　**c.**　　　　**Check**

$$\begin{array}{r} 430 \\ \times\ 852 \\ \hline \end{array}$$ 　　　　$$\begin{array}{r} 3{,}725 \\ \times\ 612 \\ \hline \end{array}$$ 　　　　$$\begin{array}{r} 8{,}008 \\ \times\ 907 \\ \hline \end{array}$$

3. a.　　　　**Check**　　**b.**　　　　**Check**　　**c.**　　　　**Check**

$$\begin{array}{r} 4{,}500 \\ \times\ 425 \\ \hline \end{array}$$ 　　　　$$\begin{array}{r} 4{,}045 \\ \times\ 222 \\ \hline \end{array}$$ 　　　　$$\begin{array}{r} 521 \\ \times\ 424 \\ \hline \end{array}$$

4. a.   **Check**   **b.**   **Check**   **c.**   **Check**

```
    932                    654                  8,034
  × 283                  ×  29                ×  419
```

5. Here is a partial invoice. Check the invoice, and show the correct extensions and total.

Quantity	Unit Price	Extension
128 kg	$3.05/kg	$390.40
90 kg	1.19/kg	101.70
51 kg	0.51/kg	26.01
209 kg	0.81/kg	169.29
375 kg	1.10/kg	412.50
	Total	$1,099.90

Answers

5._____

_____

_____

_____

_____

_____

Estimating Answers in Multiplication

Answers to multiplication problems can be estimated by rounding the multiplier and the multiplicand, usually to the greatest digit, and then multiplying the rounded numbers. The answer obtained will be close to the actual answer. Do not round single-digit numbers.

EXAMPLE

Multiply: 432 by 29.

SOLUTION

Estimate	Actual
400	432
× 30	× 29
12,000	12,528

Self-Check Answer
Estimate 600,000
Actual 637,450

Self-Check
Estimate the product: 28,975 × 22. Then find the actual product.

Problems

Estimate the product. Then compute the actual answer.

6.

a.

Estimate	Actual
	178
	$\times\ 8$

b.

Estimate	Actual
	295
	$\times\ 19$

c.

Estimate	Actual
	1,003
	$\times\ 32$

d.

Estimate	Actual
	650
	$\times\ 21$

e.

Estimate	Actual
	994
	$\times\ 101$

f.

Estimate	Actual
	428
	$\times\ 56$

Business Applications 2.2

Answers

1. _____

2. _____

3. _____

Estimate. Then solve.

1. Wilson Painters charges $12 an hour for an exterior painter. If it takes a painter 75 hours to paint a house, what is the total labor cost of the job?

2. Hartford Services' branch manager earns $5,000 per month. What is Hartford Services' branch manager's yearly salary?

3. R & J Automotive charges $52 per hour for service. If a car requires 3 hours of service, what is the total labor cost of the job?

Activities 2.2

Graph Activity

Look at the chart below. If an apartment complex charged the average rent in Minneapolis and there were 300 units rented, how much would the manager collect each month? Estimate the answer, then calculate the exact answer. _____

Hot apartment markets

Apartment occupancy in professionally managed buildings averaged 94.9% in the first quarter, and monthly rents averaged $780. Metro areas with top occupancy rates:

	Occupancy	(average rent)
Newark, N.J.	99.4%	($711)
Middlesex, N.J.	99.1%	($798)
Minneapolis	97.4%	($951)
West Palm Beach, Fla.	97.4%	($851)
Washington, D.C.	97.0%	($941)
San Diego	97.0%	($846)

Source: M/PF Research

By Cindy Hall and Sam Ward, USA TODAY

Challenge Activity

Make a list of birthday and other special occasion gifts you will give for the year. Set a price limit for the gifts according to your budget. Estimate how much you will spend for gifts in one year's time.

Internet Activity

Need to multiply a number by 5? Rather than reach for your calculator, try this method. Use this trick and attempt to multiply 85 by 5 by the estimation rounding method.

http://www.magicnet.net/~emil/ math/math_trk/multiply.html

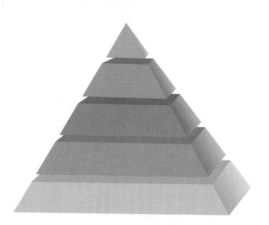

SKILLBUILDER 2.3

Using Multiplication in Business

Learning Objective

• **Solve business applications involving multiplication.**

Using Multiplication in Business

In many business applications involving multiplication, one of the factors is a money amount, written with a dollar sign and decimal point. When finding the product of an amount of money and a whole number, you must be sure to place the decimal point and dollar sign in the answer.

EXAMPLE

Determine the wholesale cost of 485 sweaters if the wholesale unit price is $15.85.

SOLUTION

```
$    15.85
×     4 85
      79 25
  1 268 0
  6 340
$7,687.25    Place a dollar sign and decimal point in the answer
```

Self-Check
Determine the wholesale cost of 500 picture frames if the wholesale unit price is $9.99.

> **MATH TIP**
> You can find the product for $3.99 × 500 mentally by multiplying $4 × 500 and subtracting $0.01 × 500:
> $4.00 × 500 = $2,000
> $0.01 × 500 = $5.
> $2000 − $5 = $1,995

Self-Check Answer
$4,995.00

Problems

Solve the following problems.

1. Midtown News Service inventory shows the following record of newspapers sold last week. What is the total of each newspaper sold and the total wholesale cost to Midtown?

Newspaper	Daily Sales							Total For Week	Wholesale Unit Price	Amount
	M	T	W	TH	F	SA	SU			
A	350	375	320	360	370	280	–0–	_____	$0.35	_____
B	500	511	485	470	462	370	–0–	_____	0.21	_____
C	185	197	221	184	170	152	–0–	_____	0.23	_____
D	212	249	248	240	225	185	–0–	_____	0.26	_____
E	408	417	441	416	409	397	–0–	_____	0.28	_____
F	160	173	181	173	168	160	–0–	_____	0.41	_____
G	–0–	–0–	–0–	–0–	–0–	–0–	402		0.97	
H	–0–	–0–	–0–	–0–	–0–	–0–	508		0.53	
									Total Cost	_____

2. Complete the following invoice.

KAPLAN'S KNITS INC.
124 Crescent Street
San Francisco, CA 94111

To The Right Fit
3728 Harrison Street
Oakland, CA 94602

Invoice Number 385
Date   May 10, 20—
TERMS 2/10, n/30
VIA   PIE

	Quantity	Item	Unit Price	Amount
a.	16	Women's knit suits	$160.00	_____
b.	18	Sweatshirts	52.00	_____
c.	24	Knit tops	12.00	_____
d.	40	Cotton knit shorts	9.75	_____
e.	30 doz	Knee socks/doz	21.50	_____
f.	180	T–shirts	8.00	_____
			Total	_____

Calculator Tip

If you use the calculator memory to store each subtotal in a calculation, you can find the total by using the memory recall feature.

Skillbuilder 2.3   Using Multiplication in Business   **83**

3. Complete the following invoice.

TALON'S CONSTRUCTION COMPANY
P.O. BOX 8729 — BELVIDERE, NJ 07823

	Item No.	Quantity	Description	Price/Unit	Extension
a.	24657	1 ea.	Hi-Performance Vac.	$ 69.99	_____
b.	Stan10	7 ea.	Stain brush—1 in.	2.59	_____
c.	214SRSC	28 box	6 × 2¼ sheet rock screws	3.82	_____
d.	501-944	7 ea.	Exterior door	169.99	_____
e.	086-441	12 ea.	Window—3.2 × 4.5	152.60	_____
f.	6-002-182	25 shts	Plywood—⅝—CDX	16.79	_____
g.	6-308-324	24 bdl	Roof shingles—brown	12.50	_____
h.	6-004-048	3 rls	Roofing felt—15 lb	10.99	_____
i.	6-114-02	50 lb	Roofing nails	1.50	_____
				Total	_____

Student Success Hints

Visual learners are those who primarily have to see something to make sense of it. Visual learners learn math by reading, doing math mentally, and using handouts.

Activities 2.3

Graph Activity

Look at the graph below. If the average grand piano sold for $10,850 in 1997, what was the approximate amount of revenue that resulted from total sales of grand pianos in that year? _____

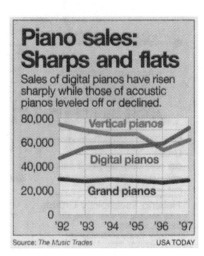

Challenge Activity

Erin drives 10 miles to work each weekday and 10 miles home. Her place of employment is directly north of her house. Her aunt lives exactly five miles south of Erin. One Monday, Erin drove to her aunt's house before going to work. The rest of the week, Erin only drove to and from work. On Saturday and Sunday, Erin stayed home. How many miles did Erin drive that week? _____

Internet Activity

A cattle ranch is not a natural ecosystem. Thus it is necessary to determine if there are enough grazeable plants for all the animals. A cow's daily energy need is called its "SUE" rate. A lactating cow, a bit over a thousand pounds, has a SUE of 1.15. To know the cow's daily meal needs, multiply the SUE by 19.6 dry digestible pounds of plants. How many pounds of plants would a lactating cow of 900 pounds need?

http://texnat.tamu.edu/pubs/b-1606/B-1606-3.htm

http://texnat.tamu.edu/pubs/b-1606/B-1606-4.htm

SKILLBUILDER 2.4

Dividing Whole Numbers

Learning Objective

• **Divide one whole number by another.**

Dividing Whole Numbers

In division, the **dividend** is divided by the **divisor.** The result of the division process is the **quotient.** If the quotient is not a whole number, the fractional part can be expressed as a **remainder (R).** In the following example, the divisor (25) divides into the dividend (76) 3 times with a remainder of 1 (R1):

Quotient ─┐ ┌─ Remainder

Divisor → 25)76 ← Dividend

3 R1

$$-75$$

1

To divide, estimate how many times the first digit in the divisor goes into the first digit of the dividend. Write this estimate in the quotient and multiply it by the divisor. Then subtract the product from the dividend, as the example illustrates.

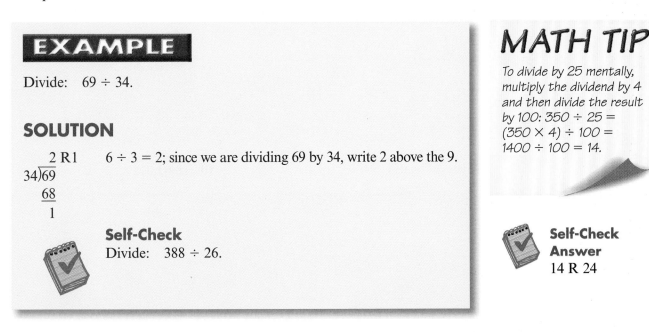

EXAMPLE

Divide: 69 ÷ 34.

SOLUTION

2 R1

34)69

68

1

6 ÷ 3 = 2; since we are dividing 69 by 34, write 2 above the 9.

Self-Check
Divide: 388 ÷ 26.

MATH TIP

To divide by 25 mentally, multiply the dividend by 4 and then divide the result by 100: 350 ÷ 25 = (350 × 4) ÷ 100 = 1400 ÷ 100 = 14.

Self-Check Answer
14 R 24

Answers

1. _____
2. _____
3. _____
4. _____
5. _____
6. _____
7. _____
8. _____
9. _____
10. _____

Find the quotient in each of the following problems. If the quotient is not a whole number, show the remainder.

1. $7\overline{)952}$

2. $4\overline{)1,684}$

3. $5\overline{)2,075}$

4. $3\overline{)2,474}$

5. $16\overline{)592}$

6. $43\overline{)9,419}$

7. $633\overline{)187,564}$

8. $458\overline{)219,158}$

9. $17\overline{)34,051}$

10. $18\overline{)63,018}$

SKILLBUILDER 2.5

Checking and Estimating With Division

Learning Objectives

* **Check the accuracy of division problems using multiplication.**
* **Estimate the answer to division problems using rounded numbers.**

Checking Division

To check the accuracy of a division computation, multiply the divisor by the quotient and add any remainder. The result should equal the dividend.

$$(\text{Divisor} \times \text{Quotient}) + \text{Remainder} = \text{Dividend}$$

EXAMPLE

Divide: 2,586 ÷ 26. Check your answer.

SOLUTION

```
      99 R12        Check
26)2,586              99
   2 34            × 26
     246             594
     234           1 98
      12           2 574
                 +    12
                   2,586
```

Self-Check
Divide: 3,427 ÷ 23. Check your answer.

> ## MATH TIP
>
> *When dividing whole numbers, the quotient will be less than the dividend unless the divisor is 1.*

Self-Check Answer
149

Problems

Find the quotient in each problem, and indicate what the remainder is, if any. Check your answers by the method shown in the preceding example. Show the details of your work.

                                Check   **Check**

1. $62\overline{)3{,}658}$ **2.** $38\overline{)5{,}780}$

                                Check   **Check**

3. $84\overline{)9{,}660}$ **4.** $75\overline{)22{,}726}$

                                Check   **Check**

5. $5\overline{)251{,}530}$ **6.** $42\overline{)11{,}258}$

                                Check   **Check**

7. $48\overline{)3{,}985}$ **8.** $77\overline{)15{,}864}$

Name:_____ Date: _____

Estimating Answers in Division

Answers to division problems can be estimated by rounding the divisor and the dividend, usually to the first digit. It is not necessary to round the divisor if it is a single-digit number.

When the second digit of the number is 5 or more, it is usually more accurate to round off to the second digit instead of the first. For example, you can more accurately estimate 17,784 ÷ 152 by finding 18,000 ÷ 150. It is also helpful to use *compatible* numbers—numbers that can be divided easily. For example, to estimate 14,124 ÷ 428, use 12,000 ÷ 400, because 12 and 4 are compatible.

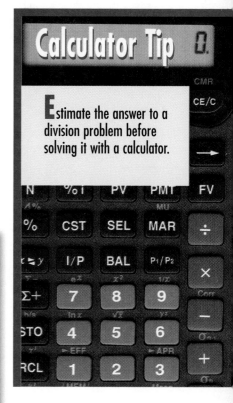

Calculator Tip

Estimate the answer to a division problem before solving it with a calculator.

Estimate: 6,192 ÷ 48. Find the actual answer.

SOLUTION

Actual	Estimate
129	120
48)6,192	50)6,000
4 8	
1 39	
96	
432	
432	

Self-Check
Estimate: 960 ÷ 8. Then find the actual answer.

Self-Check Answer
Estimate 100
Actual 120

Estimate the answers. Then find the actual answers. Show the details of your work.

9. Estimate   **Actual**   **10. Estimate**   **Actual**

                  42)8,294   28)92,540

Business Applications 2.5

1. The Ivery Financial Investment Company sent 23 employees to a three-day seminar. If the total cost was $18,975, estimate the cost per employee. Then find the actual costs.

2. The total weekly salaries of 12 cashiers at Sweet's Supermarkets are $5,280. Estimate each cashier's earnings for the week. Then find the actual salaries.

Name:_____ Date:_____

Activities 2.5

Graph Activities

Approximately how many times larger is the United States' investment and asset management industry's net revenue than Brazil's? Round to the nearest whole number. _____

Investment superpowers

The investment and asset management industry's global net revenue[1] totaled $277 billion in 1996 (latest available) and is expected to be $900 billion by 2010. Countries with the largest share, in billions, of the 1996 total:

USA $138
Japan $21
Britain $14
France $12
Germany $11
Canada $11
Brazil $9

1 – Total revenue minus costs earned by mutual funds, unit investment trusts, pensions, private banking, retail securities brokerages, private equity

Source: Economist Intelligence Unit   By Anne R. Carey and Dave Merrill, USA TODAY

Challenge Activity

Ms. Zaudke is a lawn care specialist. Mr. Harrod has contracted her to winterize his lawn. She will apply fertilizer at one (1) bag per 1,000 square feet (To find the total square footage of the lawn use this formula—**Length × Width = Area**). You are Ms. Zaudke's assistant and your challenge is to measure any rectangular or square lawn (this will be Mr. Harrod's lawn). From these measurements determine how many bags of fertilizer to apply.

   If Ms. Zaudke charges $25.00 per 1,000 sq. ft., how much will Mr. Harrod pay to winterize his lawn?

Internet Activity

Use the weight chart at **http://www.adventureteam.com/tricksofthetrail.html** to determine what it will take to carry water on your next backpacking trip. There will be three people on the hike, and you will be taking two gallons of water. How much of a load will it be at the start of the hike for each person, if the water is divided equally?

SKILLBUILDER 2.6

Using Division in Business

Learning Objective

* Solve business applications involving division.

Using Division in Business

Many products are priced and sold in units of 100 or 1,000. The abbreviation for 100 units is C, and the abbreviation for 1,000 units is M. A hundredweight, or 100 lb, is abbreviated cwt. The abbreviation for 1,000 board feet of lumber is MB or MBF. Other abbreviations you may encounter are doz (dozen) and gr (gross, or 144 pieces).

EXAMPLE

An office supply store sells envelopes at $2.50/C. Find the cost of 600 of these envelopes.

SOLUTION

First we need to determine the number of 100 units in 600 envelopes:

$$\frac{600}{100} = 6$$

600 envelopes at $2.20/C → 6 × $2.50 = $15.00
The envelopes cost $15.00

MATH TIP

To divide by powers of 10 (10, 100, 1,000, etc.), drop the same number of digits, starting at the right, as you have zeros in the power of 10. The dropped digits will be the remainder. For example: 9,427 ÷ 10 = 942 R7 and 35,600 ÷ 100 = 356 (R0).

Self-Check
Talon's Lumber Yard bought 8,000 board feet of lumber for $27/MBF. What was the cost of the lumber?

Self-Check Answer
$216

Answers

1. _____

2. _____

3. _____

Problems

Solve the following problems.

1. A grocery supplier sold 900 lb of sugar for $33.50/cwt. What did the supplier charge for the sugar?

2. Clint's Lawn Service bought 8,000 kg of potash for $159/M kg. What was the charge listed on the invoice?

3. A nursery ordered 720 packets of flower seeds priced $57.60/gr. What was the cost of the flower-seed packets?

Complete the extensions on each of the following invoices. Round to the nearest cent, where necessary.

Student Success Hints

Auditory learners primarily have to hear something to understand it. Auditory learners can learn math by listening to others, asking questions, and discussing problems with tutors and/or study groups.

4.

GLADEAU GARDEN SUPPLY
212 State Highway 30
Hohokus, NJ 07423

To The Green Thumb
1123 East End Road
Weehawken, NJ 07087

Invoice Number 11857
Date March 3, 20—
TERMS C.O.D.
VIA Pickup

	Quantity	Item	Unit Price	Amount
a.	8,000 kg	# 18–216 Weed block	$909/M kg	_____
b.	1,200 lb	# 18–128 Green seed	133.33/cwt	_____
c.	800 bags	# 18–861 Peat grow— 40 lb	80.00/C bags	_____
d.	3,000 yd	# 18–887 Burlap	970/M yd	_____
			Total	_____

Name:_____ Date:_____

5.

MILTON PAPER SUPPLY
843 Ridgedale Avenue
Cedar Knolls, PA 18350

			Invoice Number 12289
To	Rainbow Caterers	**Date**	May 1, 20—
	4298 Mt Pleasant Avenue	**TERMS**	Net
	Livingwell, NJ 07793	**VIA**	Delivered

	Item	Quantity	Description	Unit Price	Extension
a.	11–852	10 M	Napkins—economy	$ 3.71/M	_____
b.	41–725	250 C	Plates—5 in.	2.87/C	_____
c.	41–730	110 C	Plates—10 in.	5.21/C	_____
d.	58–0710	480 doz	Cutlery sets	21.60/gr	_____
e.	29–365	5 L	Liquid Kleen concentrate	2.89/L	_____
				Total	_____

6.

M. SWITZER AND SON INC.
2015 Jersey Avenue
Euclid, OH 44120

			Invoice Number 1285
To	Euclid Art Supplies	**Date**	February 25, 20—
	4285 Race Street	**TERMS**	Net
	Euclid, OH 44120	**VIA**	Picked up

	Quantity	Item	Unit Price	Amount
a.	18 rolls	Nu–Felt—50 yd/roll	$ 45.00/C yd	_____
b.	36 doz	Watercolor sets	338.40/gr	_____
c.	25,000 shts	Kraft paper—assorted colors	19.50/M	_____
d.	500 kg	Sculpt-clay	418/100 kg	_____
e.	2,000 sq ft	Canvas board—assorted sizes	369/M sq ft	_____
			Total	_____

Activities 2.6

Graph Activities

How many people per square mile live in Hawaii? Round your answer to the nearest whole number. _____

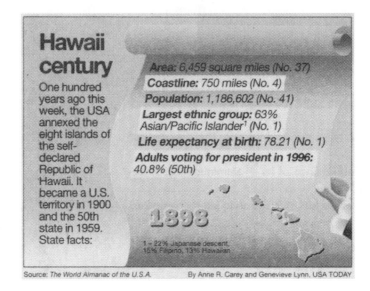

Hawaii century

One hundred years ago this week, the USA annexed the eight islands of the self-declared Republic of Hawaii. It became a U.S. territory in 1900 and the 50th state in 1959. State facts:

Area: 6,459 square miles (No. 37)
Coastline: 750 miles (No. 4)
Population: 1,186,602 (No. 41)
Largest ethnic group: 63% Asian/Pacific Islander[1] (No. 1)
Life expectancy at birth: 78.21 (No. 1)
Adults voting for president in 1996: 40.8% (50th)

1898

1 – 22% Japanese descent, 15% Filipino, 13% Hawaiian

Source: The World Almanac of the U.S.A.   By Anne R. Carey and Genevieve Lynn, USA TODAY

Challenge Activity

Ask a travel agency for tour books that explain various tours and the cost of each. Plan a trip with a friend. Determine the total price of the trip and what it will cost each of you to take this trip. Consider tips and taxi fares.

Internet Activity

The 100-year old Dow Jones industrial average uses a divisor of 0.33839549. Why is that? Find out what an unweighted average is, and why a high-market value stock gets the same weight as a low-market value stock.

http://www.phillynews.com/online/finance/prs53096.html

SKILLBUILDER 2.7

Computing Averages

Learning Objective

• **Compute the average of a group of numbers.**

Computing Averages

The **average** of a group of numbers is found by dividing the sum of the set of numbers by the number of items in the set. The average is also called the **mean** of the numbers.

EXAMPLE

During a 3-year period, a sales representative earned $28,900; $32,450; and $38,526. Find the average annual income for this period.

SOLUTION

$28,900
 32,450
 38,526
$99,876   Total 3-year income

$99,876 ÷ 3 = $33,292   Average annual income

> **MATH TIP**
>
> When computing averages, the solution is reasonable if it is between the lowest and highest amounts being added.

Self-Check
Initial investments in businesses in Coulterville were $36,645; $51,000; $44,500; $50,125; and $55,300. Find the average investment.

Self-Check Answer
$47,514

Problems

Solve the following problems.

1. During this year's United Campaign, 112 communities in Morris County contributed a total of $4,974,144. What was the average contribution per community?

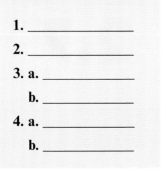
2. Four sales representatives employed by Randolph Realty Associates earned $4,220; $2,862; $2,640; and $5,150 during the past month. What was the average of their monthly earnings?

3. Marge Pineros is a sales representative with Randolph Realty Associates. For the past 12 months, her earnings were $3,364; $3,695; $3,215; $4,150; $2,915; $4,197; $3,496; $3,751; $3,215; $3,935; $3,202; and $3,525.
   a. What were her total earnings for the year?
   b. What were her average monthly earnings to the nearest cent?

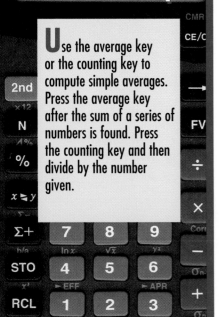
4. The nine sales representatives for Bledsoe Products had expense accounts for the past 3 months as follows: $1,128; $1,374; $801; $1,506; $1,251; $888; $1,029; $1,317; and $1,533.
   a. What was the average expense account for the 3-month period?
   b. What was the average monthly expense account?

5. Stella Chandler uses her car for business. For the past year, she drove the following mileage. Write the cost of gasoline per quarter and the year in the Total Cost column.

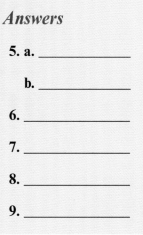

Answers

5. a. _____

   b. _____

6. _____

7. _____

8. _____

9. _____

Quarter	Miles	Gallons of Gas Used	Price Per Gallon	Total Cost
Jan.–Mar.	1,806	111	$1.23	_____
Apr.–June	3,015	196	$1.22	_____
July–Sept.	3,674	240	$1.20	_____
Oct.–Dec.	2,870	163	$1.21	_____
Total	_____	_____	_____	_____

   a. What was her average mileage for a gallon of gas? Round your answer to the nearest hundredth.

   b. What was the average price she paid for a gallon? Round off your answer to the nearest cent.

6. A jeweler sold the following numbers of rings each week during a recent month: 93, 84, 90, 77. What was the average number of rings sold each week?

7. The weekly commissions for Greg Restik, a furniture salesperson, for the past 4 weeks were $1,000, $850, $1,200, and $900. What was his average commission per week?

8. Naomi Melesco, a produce auctioneer, took a total of 3,000 bags of potatoes to auction. Of the 3,000 bags, 40 bags were spoiled and were dumped prior to the sale. 960 bags were sold for $6.75 per bag; 1,100 bags for $6.25 per bag; 500 bags for $7.15 per bag; and the remainder for $7.10 per bag. What was the average price per bag?

9. Long's Lilies bought 3,500 planters for $3,500 at a bankruptcy auction. It sold 1,850 planters for $2.70 each, and 1,000 for $2.85 each. For how much would Long's Lilies have to sell the balance (to the nearest dollar) in order to realize the average selling price of $3 a planter?

Activities 2.7

Graph Activity

Look at the chart below. How many pages of the GATT does it take to weigh a pound? _____

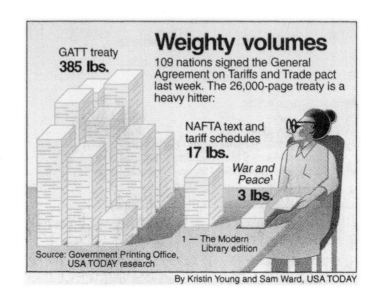

Weighty volumes

GATT treaty
385 lbs.

109 nations signed the General Agreement on Tariffs and Trade pact last week. The 26,000-page treaty is a heavy hitter:

NAFTA text and tariff schedules
17 lbs.

War and Peace[1]
3 lbs.

1 — The Modern Library edition

Source: Government Printing Office, USA TODAY research

By Kristin Young and Sam Ward, USA TODAY

Challenge Activity

Walters Construction built five model homes. The prices of four of the models were $100,000, $108,000, $126,000, and $140,000. If the mean price of the five models is $128,800, what is the price of the fifth model?

_____

Internet Activity

What is the average U.S. national unemployment rate? How does your state compare? See what the U.S. Bureau of Labor Statistics has to report, at

http://www.dismal.com/toolbox/ dict_employ.stm.

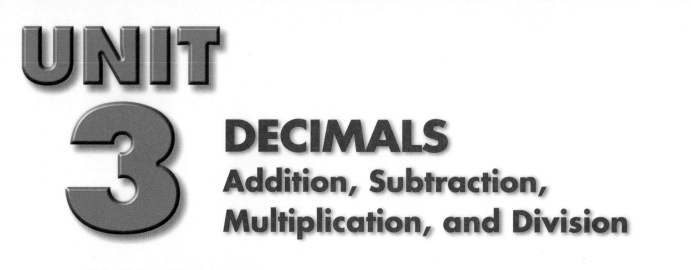

UNIT 3

DECIMALS
Addition, Subtraction, Multiplication, and Division

The chart shown here gives total expenditures for advertisers by category and medium. The expenditures are given in decimals, in millions of dollars. Why do you think this is the case? Which medium attracted the most advertising? In which category was the most money spent? What factors might influence the media in which a company would choose to advertise?

In this unit we study operations with decimal numbers, as well as rounding decimals and shortcuts in multiplying and dividing decimals.

				TOTAL U.S. AD SPENDING BY CATEGORY AND MEDIUM					
Category	Total Ad spending	Magazines	Sunday magazines	Newspaper	Network TV	Spot TV	Syndicated TV	Cable TV	Network radio
Automotive	$5,259.1	$940.3	$38.3	$746.2	$1,559.5	$1,480.6	$120.7	$117.7	$72.2
Retail	5,125.9	197.1	93.0	2,491.7	351.7	1,528.0	40.1	42.9	92.1
Business, consumer serv.	3,582.8	444.3	28.4	1,207.2	598.3	811.2	72.8	112.2	77.2
Food	3,551.9	433.4	29.1	36.5	1,443.8	974.0	346.1	141.0	41.9
Entertainment	2,913.2	58.5	36.7	387.3	802.3	1,196.0	196.1	83.0	15.6
Toiletries & cosmetics	2,249.3	631.8	21.9	6.8	958.6	293.3	215.3	92.5	8.2
Travel & hotels	2,123.2	335.4	49.1	1,064.5	212.7	233.6	8.6	49.1	33.4
Drugs & remedies	1,807.9	164.8	19.5	78.2	851.6	341.4	171.3	80.2	56.3
Direct response cos.	1,192.9	555.7	302.3	65.1	39.7	76.6	33.6	53.2	62.1
Candy, snacks & soft drinks	1,146.2	55.2	2.4	13.5	434.3	318.3	185.1	72.2	20.5
Apparel, footwear	904.8	416.0	21.5	7.0	235.4	88.5	58.9	41.0	7.5
Beer & wine	839.6	51.6	5.4	12.8	347.0	209.2	60.1	43.5	1.5
Insurance & real estate	705.3	141.6	8.9	187.3	171.5	106.2	9.1	19.0	19.1
Publishing & media	700.4	195.7	11.8	194.1	40.3	137.7	12.1	19.1	21.9

MATH CONNECTIONS

Supermarket Manager

A supermarket manager supervises the employees, plans schedules, places food orders, and sets prices. Each of these functions is important to the store's success. The decisions made in hiring, firing, and scheduling are basically a matter of the manager using good judgment based on experience. However, to order food to stock shelves and to set prices for all the items, a manager must use math skills. Knowing the competition and understanding manufacturers' requirements are essential.

Math Application

A store manager wants to estimate the weekly cost of meat for a family of 2 adults and 2 children for next week's ad. He chooses 6 pounds of chicken ($2.49/lb), a 5-pound pork roast ($2.89/lb), one package of hot dogs ($1.49/pkg), and 3 cans of tuna ($.85/can). What is the manager's estimate?

Critical Thinking Problem

The store manager decides to sell a gallon of milk for $.80 below cost in order to lure customers into the store. What are some ways that the manager might try to regain the loss on the milk?

SKILLBUILDER 3.1
Reading, Writing, and Rounding Decimals

Learning Objectives

- **Read and write decimals using words.**
- **Read word names of decimals and write decimal names using digits.**
- **Round decimals to a specified place.**

Reading and Writing Decimals

A **decimal number,** or **decimal,** is a number containing a decimal point and digits to the right of the decimal. The digits to the right of the decimal point are called **decimals.** Decimals have a value of less than one. Each place in a decimal is one tenth the value of the place to its left.

A digit in the hundredths place has a value of one-tenth of the same digit in the tenths place. A digit in the thousandths place has a value of one-tenth of the same digit in the hundredths place.

To read or write a decimal in words, read the digits as if they were a whole number and then read the name of the place of the digit on the extreme right. If the decimal has a whole number part, read the number to the left of the decimal point. Next, read the decimal point as *and.* Finally, read the name of the decimal.

Look at the Example and Solution shown on the next page.

MATH TIP

Do not use the word and when reading the whole number part—and is read only when a decimal point occurs.

Read 203.167 and then write it in words.

SOLUTION

Write the number in a place-value chart.

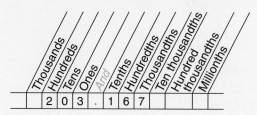

So, 203.167 is read "two hundred three and one hundred sixty-seven thousandths."

Self-Check Answer

Five thousand eight hundred ninety-two ten thousandths

Self-Check

Read the number 0.5892 and write it in words.

Name:_____ Date: _____

Problems

Read these numbers and then write them in words.

1. 0.7   _____
2. 0.23   _____
3. 0.68   _____
4. 4.016   _____
5. 5.118   _____
6. 17.83   _____
7. 24.216   _____
8. 3.0808   _____
9. 25.2006 _____
10. 4.00006 _____
11. 379.058 _____
12. 8,016.09 _____

Student Success Hints

Remember to place commas properly when writing decimals. Commas appear only on the left side of the decimal point. No commas are used on the right side of the decimal point. Write two thousand, eight hundred six and one thousand five hundred forty-two ten thousandths as 2,806.1542.

Writing Decimals Using Digits

A place-value chart can also be used to help write the word name of a decimal using digits.

EXAMPLE

Write the number four hundred five ten thousandths using digits.

SOLUTION

Write the number 405 in a place-value chart, placing the 5 in the ten thousandths place. Zeros must be added to any empty places between the number and the decimal point.

Tens	Ones	And	Tenths	Hundredths	Thousandths	Ten thousandths	Hundred thousandths	Millionths
	0	.	0	4	0	5		

Four hundred five ten thousandths is written 0.0405. Notice that we write a 0 in the units place when writing decimal fractions.

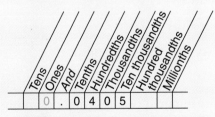

Self-Check

Read the number sixteen hundredths and write it using digits.

Student Success Hints

If you are having difficulty reading or writing decimals, make up numbers and use the place-value chart on this page. To help you, make several blank charts on your paper and place your numbers on the chart. Now read or write the number. Repeated practice will help you learn and remember.

Self-Check Answer
0.16

Problems

Read these numbers, and then write them using digits.

Answers

13. _____

14. _____

15. _____

16. _____

17. _____

18. _____

19. _____

20. _____

21. _____

22. _____

13. Three hundred seventy-three thousandths

14. Eighty-four ten thousandths

15. Forty-seven hundredths

16. Seven tenths

17. Thirty-one and six thousandths

18. One thousand, sixty and six tenths

19. Twenty-seven and thirteen hundredths

20. Three and nine hundred nine ten thousandths

21. One and sixteen thousandths

22. One and nine hundred thousandths

Write the following numbers using digits in the columns at the right. Align them correctly both horizontally and vertically.

23. a.

b.

c.

d.

e.

23. **a.** One hundred sixteen

b. Five and six thousandths

c. Thirty-nine

d. Fifty-five and one hundred fifty-one thousandths

e. Eighty-nine and seven tenths

24. a.

b.

c.

d.

e.

24. **a.** Two thousand, sixty three and four tenths

b. Sixty-six and sixty-six hundredths

c. Thirty-five and five tenths

d. Six thousand, three hundred twenty-seven

e. Five hundred fourteen

25. a.

b.

c.

d.

e.

25. **a** Twenty-seven and three hundred fourteen ten thousandths

b. Seventeen and seventeen thousandths

c. Forty-seven and forty-seven hundredths

d. Nine thousand seven hundred sixty-two ten thousandths

e. Twenty-nine and eight tenths

26. a.

b.

c.

d.

e.

26. **a.** Nine hundred seventy-two and one hundred twenty-five thousandths

b. Forty-three and forty-three thousandths

c. Fifty-two and five tenths

d. Three and three thousandths

e. Seven hundred six and sixty-seven thousandths

Rounding Decimals

Rounding decimals to a specified place is very similar to rounding whole numbers. To round a number to a specified place, identify the digit in the rounding place. Then look at the digit to the right of that place. If it is 5 or more, increase the digit in the rounding place by 1 and drop the digits to the right of the rounding place. If the digit is less than 5, drop the digits to the right of the rounding place.

EXAMPLE

a. Round 2.7063 to the nearest tenth.
b. Round $104.725 to the nearest cent.

SOLUTION

a. Place a line under the digit in the tenths place. Then place an arrow over the digit to the right.

$$\downarrow$$

2.7063

Since 0 is less than 5, drop the digits to the right of 7.

2.7063 to the nearest tenth is 2.7.

b. Place a line under the digit in the cents place. Then place an arrow over the digit to the right.

$$\downarrow$$

$104.725

Since 5 is 5 or greater, increase the cents digit by 1 and drop the 5.

$104.725 to the nearest cent is $104.73.

Self-Check
Round $39.6027 to the nearest cent.

Calculator Tip

Most calculators with rounding selectors are set on the 5/4 position except when finding unit price (any fraction of a cent is passed on to the consumer).

Self-Check Answer
$39.60

Problems

Round to the specified place.

27. 53.79, nearest tenth _____

28. 0.9432, nearest thousandth _____

29. $16.827, nearest cent _____

30. $12.695, nearest cent _____

31. 8.995, nearest hundredth _____

32. $0.864, nearest cent _____

33. 2,756.8051, nearest hundredth _____

34. $1.555, nearest cent _____

Business Applications 3.1

1. a. _____

   b. _____

   c. _____

   d. _____

   e. _____

   f. _____

   g. _____

   h. _____

   i. _____

   j. _____

2. _____

1. Kelso Enterprises sells merchandise by catalog. They include a chart that shows the shipping charges, which are determined by the dollar amount of the order. Use the table to determine the shipping charge for each order.

SHIPPING AND HANDLING	
Order Amount	**Shipping Charge**
Up to $15.00	$3.75
$15.01 – $30.00	$5.35
$30.01 – $50.00	$6.35
$50.01 – $80.00	$6.95
$80.01–$100.00	$7.55
$100.01–$200.00	$7.95
Orders over $200.00	$8.25

a. $ 8.25

b. $ 45.00

c. $ 73.40

d. $198.00

e. $ 26.30

f. $505.34

g. $ 97.50

h. $ 18.00

i. $201.00

j. $ 30.02

2. Shelves and More computed the average cost of a shelf unit to be $284.7961. What is the average cost to the nearest cent?

Activities 3.1

Graph Activity

To the nearest million, how many dollars of strawberry and grape jellies are sold in the USA in one year? To the nearest million, how many dollars of all other jellies are sold in the USA in one year? _____

_____

Berry popular jelly

$366.2 million

$291.5 million

Two berries account for more than half of the USA's yearly jelly purchases:

STRAWBERRY AND GRAPE

ALL OTHERS

Source: Nielsen Marketing Research

By John Riley and Bob Laird, USA TODAY

Challenge Activity

For this project, as a shopper for Tammie's Personal Shopping Service, you are to go to a local department store. Your challenge is to select five (5) items on sale. Determine how much money you saved on each item, figure the percentage saved for each item. Calculate your total purchase price and the total amount saved.

Original price	Sale price	Amount saved	Percentage saved
1.			
2.			
3.			
4.			
5.			
TOTAL _____	_____	_____	_____

Internet Activity

Foreign exchange rates are calculated each day. You can find out how one country's currency compares with another's. Practice reading decimals by finding out how the Belgian Franc compares to the Chinese Yuan. Round decimals to the hundredths place. Check back to this site in two weeks. Has the currency rate changed?

http://www.xe.net/currency/table.htm

Name:_____ Date:_____

SKILLBUILDER 3.2
Adding Decimals

Learning Objective

- **Add two or more decimals.**

Adding Decimals

When adding decimals, place the addends in a column with the decimal points aligned. Recall that annexing zeros to the right of a decimal or to the right of the decimal point in a whole number does not change the value of the decimal. It may be helpful to annex zeros to help align columns properly. Remember to place the decimal point in the sum directly below the decimal points in the addends.

EXAMPLE

Find the sum:
425.04 + 22 + 9.132 + 0.37.

SOLUTION

```
   1    1
 425.040  ←——  Annex zeros
  22.000  ←
   9.132
   0.370  ←
 456.542
```

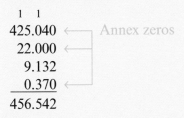

Self-Check
Find the sum: 0.34 + 23.901 + 634 + 0.8

MATH TIP

The decimal point is understood in a whole number; therefore, the decimal point does not have to be written. Remember, however, to put the decimal point to the right of the number when solving problems.

Self-Check Answer
659.041

Problems

Find each sum.

1. 621.88
    + 413.93

2. 28.51
    + 3.18

3. 46.94
    + 31.27

4. 5.79
    + 266.66

5. $14.35
    + 25.99

6. $545.10
    + 22.60

7. 7,190.9
    15.56
    0.042
    + 123.25

8. 61.188
    115.75
    0.0125
    + 9.9

9. 17.471
    36.31
    257.225
    + 15.9

10. $5,127.75
    73
    125.25
    + 2,800.00

11. $317
    12.26
    416.97
    + 9.33

12. $ 948.31
    130.40
    3,289.14
    + 51.76

Answers

13. _____

14. _____

15. _____

16. _____

13. 37.83 + 115.7 + 25.678

14. 52.15 + 74.716 + 1.12

15. 14.7 + 6.017 + 83.49 + 9

16. $75.34 + $13.50 + $215.49 + $5.00

17. 14.5 + 56.78 + 19.316 + 10

18. 4.971 + 24.72 + 220 + 1.003

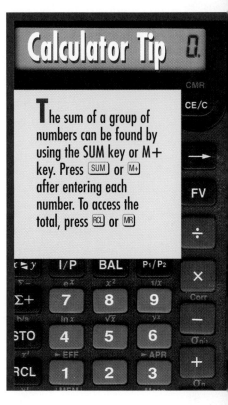

Calculator Tip

The sum of a group of numbers can be found by using the SUM key or M+ key. Press SUM or M+ after entering each number. To access the total, press RCL or MR

19. 13 + 127.7 + 263.4176 + 17.22

20. $415.27 + $51.49 + $12.45 + $1,243.00

Business Applications 3.2

1. Complete the following invoice.

Date	5/17		Total Amount	
Sales Number	005444			
Salesperson	J. Smith			

Quantity	Stock No.	Description	Unit	Amount
3 boxes	372-A	Syringes (1 doz.)	$10.25	_____
4 boxes	564-C	Cotton balls	$10.68	_____
2 boxes	119-B	$2\frac{1}{2}$-in. elastic bandages	$12.34	_____
			Total	_____

2. St. Mary's Hospital and Medical Center sent out the following bill. Compute the total due.

Patient	Doris Cunningham

Service Description	Charges
Pharmacy	$ 50.90
Laboratory pathology	41.00
Operating room	658.00
Anesthesiology	321.00
Respiratory therapy	17.00
Recovery room	155.00
Electrocardiogram	56.00
Total	_____

3. Aaron Goldman recorded his expenses on a business trip for Alco Electronics. Complete his travel expense report.

Expense Item	Su 4/27	M 4/28	T 4/29	W 4/30	Th 5/1	F 5/2	Sa 5/3	Totals
Meals			$32.00	$32.00	$32.00	$32.00		_____
Mileage			35.00					_____
Lodging			97.34	97.34	97.34	97.34		_____
Tips			6.00	7.00	6.00	7.00		_____
Taxi			17.00			17.00		_____
Tolls			1.69			1.69		_____
Airfare			212.00					_____
Parking						10.00		_____
Daily totals			_____	_____	_____	_____		_____

4. Sam stopped at Handyway on his way home from work. He bought milk for $2.99, butter for $1.74, cheese for $3.99 and eggs for $2.05. How much money did Sam spend?

Answers

4._____

5._____

5. Greta's paychecks for March were $442.30, $480.90, $400.60, and $388. What was the total of Greta's paychecks for March?

Activities 3.2

Graph Activity

If New York City gets $48.9 billion federal dollars in 1998 and $52.3 billion in 1999, how much will it have received over this three-year period (1997–1999)? _____

Counties with clout

Counties getting the most 1997 federal dollars (in billions):

$42.2 $41.4 $23.5 $17.2 $13.6

New York City   Los Angeles County   Cook County (Chicago)   San Diego County   Harris County (Houston)

Source: Census Bureau

By Cindy Hall and Marcy E. Mullins, USA TODAY

Challenge Activity

As an associate of Brittain's Insurance Co. you are to keep a record of the long distance business phone calls for quarterly reimbursement. Use your phone bill during a 3 month period and record your long distance calls as these business calls. What will be the amount of your reimbursement check?

Internet Activity

As you know, there are 365 days in a year—here on earth. (Actually, to be exact, 365.26 days.) Suppose that you were in the used space-shuttle business and spent one year selling on Mercury, one year on Mars, and one year on Venus. Using data from

http://www.uen.org/utahlink/lp_res/TRB048.html,

how many days would you be in business on other planets?

SKILLBUILDER 3.3

Subtracting Decimals

Learning Objective

• **Find the difference of two decimals.**

Subtracting Decimals

To find the difference between two decimals, write the smaller decimal below the greater one, lining up the decimal points. Annex zeros as needed so that each decimal has the same number of places. Remember to write the decimal point in the difference directly below the decimal points in the problem.

EXAMPLE

Find the difference: $156 - 1.799$.

SOLUTION

$$
\begin{array}{r}
9\ 9 \\
5\ \cancel{10}\cancel{10}\cancel{10} \\
15\cancel{6}.\cancel{0}\cancel{0}\cancel{0} \quad \leftarrow \text{Annex zeros so that 156 has three decimal places} \\
\underline{1.799} \\
154.201
\end{array}
$$

 Self-Check
Find the difference: $38.47 - 0.233$.

MATH TIP

Parentheses are generally used to indicate negative numbers in business. Thus, a loss of $2,000 is written ($2,000).

 **Self-Check Answer**
38.237

Problems

Find the difference.

1. 593. − 78.63	**2.** 437.712 − 221.353	**3.** 848.40 − 78.216
4. $88.60 − 71.70	**5.** $124.95 − 86.78	**6.** $349.57 − 128.35
7. 856.7 − 46.523	**8.** 76. − 0.437	**9.** 401. − 9.631
10. $936.00 − 341.49	**11.** $2,679.00 − 1,538.17	**12.** $3,213.56 − 1,479.57

13. 141.21 − 69.90   **14.** 246.38 − 78.83

15. 161.2 − 66.3   **16.** 721.9 − 88.33

17. 441.4 − 215.9   **18.** 44.27 − 0.6124

19. $621.66 − $78.43   **20.** $7,000 − $4,833.63

Answers

13. _____

14. _____

15. _____

16. _____

17. _____

18. _____

19. _____

20. _____

Business Applications 3.3

1. Complete the following deposit slip.

Currency	$352.00
Coins	3.75
Checks (list separately)	33.40
	129.25
	18.99
Subtotal	_____
Less cash received	$ 50.00
Total deposit	_____

2. St. Mary's Hospital and Medical Center sent a statement to Brad Hutchens following his surgery. Complete the computations to give the current balance. The patient portion is any amount remaining after insurance payments and adjustments.

St. Mary's **Hospital and Medical Center** **Patient** Brad Hutchens		**Account Number** 4302719 **Statement Date** 5/17/—

Date	Description of Service	Amount
7/17/—	Beginning balance	$1,298.90
8/07/—	Insurance adjustment	194.98
8/07/—	Payment/Insurance company	883.94
	Current balance	_____
	Patient portion	_____
	Total account balance	_____

3. Hal's Fish and Tackle prepared a report showing its projected sales in comparison to the actual sales of its best-selling rod-and-reel combinations. Find each difference to show how actual sales compare with projected sales.

HAL'S FISH AND TACKLE

Product	Sales Projection	Actual Sales	Difference
Sterling Pro 5	$3,187.15	$3,269.09	_____
Daiwa Pro Flex	$3,724.41	$3,794.35	_____
Cardinal Custom Lite	$2,146.63	$2,226.58	_____
Abu Garcia Custom Lite	$3,129.52	$3,194.46	_____
Eagle Claw Big Water	$3,841.21	$3,881.20	_____
Shimano TX/All Pro	$2,872.33	$2,919.21	_____

4. Charles went to the fair with $44.50 in his wallet. He spent $20.50 on rides and $19.89 on food. How much money did he have left?

Answers

4. _____

5. _____

6. _____

5. The petty cash fund had a beginning balance of $200. Withdrawals for the week were $12.50, $8.76, $2.50, $7.45, and $18.75. What is the balance in petty cash at the end of the week?

6. What is the difference between the dollar amount of trade with Mexico in 1993 and 1980?

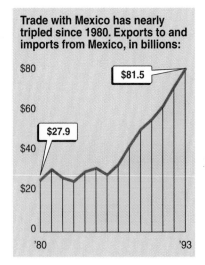

Trade with Mexico has nearly tripled since 1980. Exports to and imports from Mexico, in billions:

$81.5

$27.9

© Glencoe/McGraw-Hill

Activities 3.3

Graph Activity

Find the difference between the highest and lowest batting average.

_____

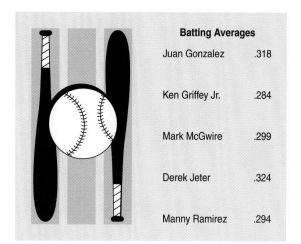

Batting Averages	
Juan Gonzalez	.318
Ken Griffey Jr.	.284
Mark McGwire	.299
Derek Jeter	.324
Manny Ramirez	.294

Challenge Activity

As an interior design consultant, Christine must determine the number of rolls of wallpaper it will take to wallpaper a room. She must follow these steps:

* Measure the perimeter of the room length × 2 + room width × 2 (round to the next highest foot).
* Measure the height of the ceiling (from floor to ceiling)
* Take perimeter × ceiling height
* This = square footage of the room's wall space; then deduct for square footage of doors, windows, and any other surface area that will not be wallpapered.
* To this square footage add another 10% to 20% (depending on paper pattern repeat). This is the total square footage for pattern match and waste.

If a single roll of wallpaper will cover 30 sq. ft. and costs $20.00, how much will it cost to wallpaper the room? Your challenge is to measure a room in your home and figure how much wallpaper it will take and what the cost will be.

Internet Activity

What if you are subtracting two numbers with decimals, but you see that one number has fewer decimals than the other? How do you align them for subtraction? Find out a solution from Dr. Math at

http://www.forum.swarthmore.edu/ dr.math/problems/merce9.16.97 .html

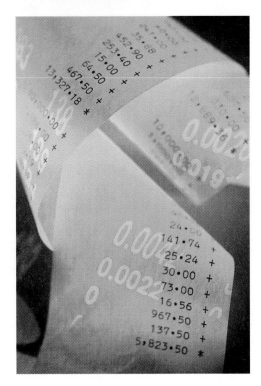

SKILLBUILDER 3.4

Multiplying Decimals

Learning Objective

• **Find the product of two decimals.**

Multiplying Decimals

Multiply decimals the same way whole numbers are multiplied. When the final product is obtained, place the decimal point so that there are as many decimal places in the answer as there are in the multiplier and multiplicand combined. If there are not enough places in the product, add zeros to the *left* of the product to make up the required number of places.

EXAMPLE

Multiply: 5.089×0.005.

SOLUTION

$$
\begin{array}{rl}
5.089 & \text{3 places} \\
\times\ 0.005 & +\ \text{3 places} \\
\hline
0.\underline{025445} & \text{Annex one zero.} \leftarrow \text{6 places}
\end{array}
$$

MATH TIP

If there are more decimal places than digits in the product, add zeros to the right of the decimal point in the product as needed.

Self-Check
Multiply: 23.06×0.25.

**Self-Check
Answer**
5.765

Answers

1. _____
2. _____
3. _____
4. _____
5. _____
6. _____
7. _____
8. _____
9. _____
10. _____
11. _____
12. _____
13. _____
14. _____
15. _____

Problems

Find the product.

1. 81.7
    × 0.57

2. 0.83
    × 0.08

3. 0.936
    × 7.381

4. $7.15
    × 65

5. $42.25
    × 3.5

6. $19.73
    × 4.3

7. 38.1 × 0.45

8. 61.7 × 3.14

9. 6.9 × 100

10. 75.5 × 0.16

11. 5.71
    × 10

12. 412.42
    × 100

13. 7.1956
    × 1,000

14. 86.3 × 10

15. $41.25 × 3.8

16. 52.751×100

17. $4.3289 \times 1,000$

18. $58.75
$\times 4.7$

19. $61.87
$\times 6.23$

20. $129.52
$\times 5.231$

21. The law firm of Anderson and Anderson employs four office personnel at an average yearly cost of $19,810 each. In order to lower office expenses, the firm has decided to purchase a word processor. With a word processor, the firm will need only two office employees, receiving an average salary of $16,310. Other employee expenses are benefits per employee, $3,500. Additional expenses with the word processor are as follows: information processing equipment, $11,495; workstation, $400; and supplies, $600. Answer the following questions.

a. Find the cost for the year with the word processor.

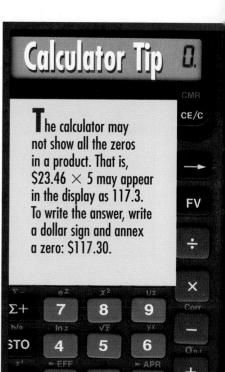

Calculator Tip

The calculator may not show all the zeros in a product. That is, $23.46 × 5 may appear in the display as 117.3. To write the answer, write a dollar sign and annex a zero: $117.30.

b. What is the yearly savings by making the change?

Business Applications 3.4

1. Arlene works for the State Department. During a recent business trip, Arlene made the following gasoline purchases: 14 gal at $1.079/gal; 18.6 gal at $.999/gal; 19.5 gal at $.959/gal; and 16.5 gal at $1.019/gal. How much did Arlene spend for gasoline on her trip? Round each purchase to the nearest cent, where necessary.

Student Success Hints

Survey all material before reading. Read the title, table of contents, objectives, introduction, headings, summary, and questions at the end of the chapter. Also look at illustrations, bolded or italicized words, and margin notes. By surveying the material, you are prepared to read more thoroughly.

2. José, the purchasing agent for Pools Etc., bought pool supplies for the business. Figure the total price of each item, the sales tax (0.06 of the subtotal), and the total cost if there is a $6 shipping fee.

Quantity	Item Number	Description	Unit Price	Total
1	82-050	Test kit	$29.99	_____
15	14-718	Oxidizer	$34.99	_____
7	14-341	Algae control	$14.99	_____
5	14-044	Alkalinity increase	$19.99	_____
5	14-045	pH increase	$24.99	_____

Subtotal	_____
Sales tax (0.06)	_____
Shipping	_____
Total	_____

© Glencoe/McGraw-Hill

Activities 3.4

Graph Activity

Suppose you bought 17.23 gallons of gasoline. According to the graph, how much would you pay? _____

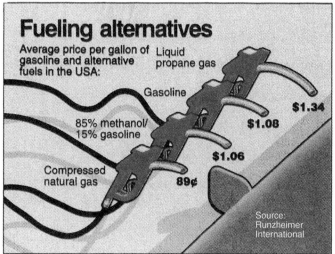

Fueling alternatives

Average price per gallon of gasoline and alternative fuels in the USA:

Liquid propane gas

Gasoline

85% methanol/ 15% gasoline

Compressed natural gas

$1.34

$1.08

$1.06

89¢

Source: Runzheimer International

By Patti Stang and Marty Baumann, USA TODAY

Challenge Activity

The distance between Tom's house and the country club is 2.25 miles and the distance between his house and the mall is 3.5 miles. Tom is on an exercise program and plans on walking 25 miles per week. If he walks on three round trips from his house to the country club, and one round trip from his house to the mall, how many miles has he left to walk?

Internet Activity

The trick in multiplying decimals, according to Dr. Math, is to disregard the decimal—at the start. What does Dr. Math mean by that? Explore the concept at

http://forum.swarthmore.edu/dr.math/problems/paul .6.27.96.html

SKILLBUILDER 3.5

Dividing Decimals

Learning Objectives

- **Divide one decimal by another.**
- **Round a quotient of decimal division to a specified place.**

Finding Quotients

When a divisor is a decimal or a mixed number, make the divisor a whole number by moving the decimal point as many places to the right as necessary. Also move the decimal point in the dividend the same number of places to the right, annexing zeros if necessary. Then divide.

EXAMPLE

Divide 62.01 by 0.018.

SOLUTION

$$\text{Divisor} \quad 0.018 \overline{)\, \underset{\text{Dividend}}{62.010}} \quad \overset{3{,}445 \;\; \text{Quotient}}{}$$

You can check the answer using multiplication: $3{,}445 \times 0.018 = 62.01$.

Self-Check
Find the quotient: $713.424 \div 200.4$.

MATH TIP

Before dividing, position the decimal point in the quotient above the new decimal point in the dividend.

Self-Check Answer
3.56

---

1. _____

2. _____

3. _____

4. _____

5. _____

6. _____

7. _____

8. _____

9. _____

10. _____

Problems

Find each quotient.

1. $4.3\overline{)45.322}$

2. $5.2\overline{)39.936}$

3. $34\overline{)456.28}$

4. $69\overline{)97.98}$

5. $21\overline{)\$121.38}$

6. $66\overline{)\$168.30}$

7. $0.1568 \div 0.49$

8. $16.7 \div 0.05$

9. $43.818 \div 6.7$

10. $3,761.8 \div 1,000$

Rounding Quotients

When a number does not divide evenly, you must decide how far to carry out the computation. This depends on the degree of accuracy required. Money amounts are usually rounded to the nearest cent. Divide one place further than the place required, and then round to that place.

EXAMPLE

Divide   49.321 ÷ 0.121. Round the quotient to the nearest hundredth.

SOLUTION

$$
\begin{array}{r}
407.611 \rightarrow 407.61 \\
0.121.\overline{)49.321.000} \\
\underline{48\ 4} \\
921 \\
\underline{847} \\
74\ 0 \\
\underline{72\ 6} \\
1\ 40 \\
\underline{1\ 21} \\
190 \\
\underline{121} \\
69
\end{array}
$$

If you check a rounded answer using multiplication, the result will not exactly equal the dividend. For example, $407.61 \times 0.121 = 49.32081$. If you round 49.32081 to the nearest thousandth, the result is the dividend, 49.321.

Self-Check
Find the quotient: 38.9 ÷ 5.44. Round the answer to the nearest thousandth.

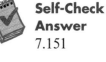

Self-Check Answer
7.151

Find the quotient. Round your answers to the nearest hundredth or the nearest cent.

Answers

11. _____

12. _____

13. _____

14. _____

15. _____

16. _____

17. _____

18. _____

11. $5.9\overline{)8{,}148.88}$

12. $64\overline{)\$728.50}$

13. $916\overline{)\$8{,}914.8}$

14. $694\overline{)\$5{,}625.97}$

15. $33.60 \div 2.1$

16. $374.4 \div 46$

Calculator Tip

Key the division problem $45 \div 5.3$ into the calculator as follows: 45 (Dividend) $\div$ 5.3 (Divisor) $=$.

17. $6.345 \div 3.9$

18. $88.91 \div 712$

Business Applications 3.5

1. Schlegel, Markham, and Johnson law firm averages the following fees for these types of cases. The firm charges $225.00 per hour. Find the number of hours spent on the average case. Round your answer to the nearest tenth.

Case Type	Fees Received	No. of Hours
a. Divorce	$23,130.00	_____
b. Bankruptcy	2,610.00	_____
c. Real estate	3,825.00	_____
d. Will preparation	990.00	_____

2. The supervisor of trucking at the Crisp Chip Company keeps a record of the fuel mileage for all trucks. Find the miles per gallon (to the nearest tenth) for these trucks in the fleet at Crisp Chip Company.

Truck	Miles	Gallons of Fuel	Miles per Gallon
A	600.0	85.7	_____
B	1,523.0	300.7	_____
C	834.5	92.6	_____
D	643.7	91.95	_____
E	2,542.6	423.766	_____

3. When a new shipment of shoes is delivered to Breymeyer Sporting Goods, the clerk in the athletics shoe department is required to compute and record the cost per pair of shoes. Find the following cost per item (to the nearest cent).

Quantity	Item	Total Cost	Cost per Item
a. 15	Running shoes	$1,039.85	_____
b. 12	Basketball shoes	1,149.49	_____
c. 20	Hiking boots	1,093.85	_____
d. 6	Baseball cleats	283.97	_____
e. 16	Wrestling shoes	517.64	_____
f. 8	Sandals	417.22	_____

Activities 3.5

Graph Activity

Look at the chart below. If members in the IMSA pay $412.95 as dues, what is the total dues paid? _____

Joining the auto racers
Membership in auto racing associations:

National Hot Rod
Association (NHRA)

National
Association for
Stock Car Auto
Racing (NASCAR)   75,000

                         42,000

International
Motor Sports
Association
(IMSA)   3,700

Championship
Auto Racing
Teams (CART)   2,000

Source: USA TODAY research   By Cindy Hall and Stephen Conley, USA TODAY

Challenge Activity

Roger, the auditor for Industrial Distributions, has decided to purchase a personal computer system for his office. As his office assistant you are to determine the cost of a computer package including the CPU, monitor, keyboard, and a printer. Your assignment is to go to a local technology store or use a computer catalog to determine the price, less rebates or any discounts, plus tax. What is the total price? Also determine what the payment schedule would be if the computer package is financed for one year. What is the difference in total cost with financing versus paying in cash?

Internet Activity

Here's a site where you can eat your cake and divide decimals at the same time:

http://www.forum.swarthmore.edu/ dr.math/problems/wright3.7.98.html

SKILLBUILDER 3.6

Shortcuts in Multiplication of Decimals

Learning Objectives

- Use shortcuts to multiply decimals by 10, 100, or 1,000.
- Use shortcuts to multiply decimals by multiples of 10.
- Use shortcuts to multiply decimals by 25, 50, 125, 250, 500, or 750.

Multiplying by 10, 100, and 1,000

To multiply by 10, move the decimal point of the multiplicand one place to the right. Add a zero if necessary.

To multiply by 100, move the decimal point two places to the right. Add zeros as necessary.

To multiply by 1,000, move the decimal point three places to the right. Add zeros as necessary.

EXAMPLE

Use the shortcut to find (a) 495 × 10, (b) 149.6 × 100, and (c) 150 × 1,000.

SOLUTION

a.   495 × 10 = 495.0   or   4,950

b. 149.6 × 100 = 149.60   or   14,960

c. 150 × 1,000 = 150.000   or   150,000

Self-Check
Use the shortcut to find 47.02 × 1,000.

MATH TIP

When multiplying by powers of 10 (such as 10, 100, or 1,000), it is easy to do the calculations mentally by moving the decimal point of the multiplicand one, two, or three places to the right.

**Self-Check
Answer**
47,020

Problems

Use the shortcuts described on page 146 to find each product. Then total each group.

1. $\$\ 875 \times\ \ \ \ 10 =$

$\ \ \ \ 36.85 \times\ \ \ 100 =$

$\ \ \ \ \ \ 8.16 \times 1{,}000 =$

$\ \ \ \ 62.50 \times\ \ \ 100 =$

$\ \ \ \ \ \ 7.95 \times 1{,}000 =$

$\ \ \ \ \ \ \ 679 \times\ \ \ \ 10 =$

$\ \ \ \ \ \ \ \ \ \ \ \ \ \ \text{Total} =$

2. $\ \ \ \ 2.36 \times\ \ \ \ 10 =$

$\ \ \ \ \ 18.1 \times\ \ \ 100 =$

$\ \ \ \ \ 9.74 \times 1{,}000 =$

$\ \ 0.6621 \times 1{,}000 =$

$\ \ \ \ 2.013 \times\ \ \ \ 10 =$

$\ \ \ \ \ 12.9 \times\ \ \ 100 =$

$\ \ \ \ \ \ \ \ \ \ \ \ \ \ \text{Total} =$

Multiplying by Multiples of 10

To multiply by a *multiple* of 10 (for example, 40, 300, or 8,000), multiply by the first digit (4, 3, or 8). Then move the decimal point of the answer the required number of places to the right.

EXAMPLE

Multiply: 746×40.

SOLUTION

$\ \ 746 \times\ \ 4 = 2{,}984$

$2{,}984 \times 10 = 2{,}984.0\ \ \ \text{or}\ \ \ 29{,}840$

**Self-Check
Answer**
3,408,000

Self-Check
Multiply: $426 \times 8{,}000$.

© Glencoe/McGraw-Hill

Use the shortcut method described to find each product. Then total each group.

Answers

3.   49 ×   50 =

        127 ×   80 =

        436 ×   300 =

         62 × 9,000 =

        919 ×   300 =

      1,585 ×   500 =

              Total =

4. 327.68 ×   400 =

      560.11 ×   60 =

      1.2222 × 4,000 =

        500 ×   50 =

        616 ×   700 =

         45 × 9,000 =

              Total =

5. $3,857 ×   50 =

        962 ×   300 =

         75 × 8,000 =

        432 ×   600 =

         56 × 3,000 =

      1,008 ×   70 =

              Total =

3._____

_____

_____

_____

_____

_____

4._____

_____

_____

_____

_____

_____

5._____

_____

_____

_____

_____

_____

Calculator Tip   0.

CMR

CE/C

Set the decimal selector on F or FL when the maximum number of decimals is needed in the product.

FV

Multiplying by 25 or by a Multiple of 25

To multiply a number by 25, first multiply by 100, then divide by 4.

To multiply by 50, multiply by 100 and divide by 2.

To multiply by 125, first multiply by 1,000 and then divide by 8.

To multiply by 250, multiply by 1,000 and divide by 4. To multiply by 500, multiply by 1,000 and divide by 2. To multiply by 750, multiply by 1,000 and divide by 4; multiply the result by 3.

EXAMPLE

Multiply: 862×25 and 963×125.

SOLUTION

$862 \times 100 = 862.00$

$86,200 \div 4 = 21,550$

$963 \times 1,000 = 963.000$

$963,000 \div 8 = 120,375$

Self-Check Answer
43,100

Self-Check
Multiply: 862×50.

Problems

Find the product in each of the following problems using the methods described.

	a.	b.	c.
6.	465×50	785×25	924×25
7.	550×125	467×500	73×250
8.	$\$675.23 \times 25$	943×50	$\$150.50 \times 250$
9.	$\$179.45 \times 50$	$\$65.75 \times 25$	$2,026 \times 125$
10.	$\$972.87 \times 125$	$\$45.38 \times 500$	$4,658 \times 250$
11.	$6,307 \times 500$	76×125	250×250
12.	375×50	$1,280 \times 125$	470×500

Answers

6. a. _____
   b. _____
   c. _____
7. a. _____
   b. _____
   c. _____
8. a. _____
   b. _____
   c. _____
9. a. _____
   b. _____
   c. _____
10. a. _____
    b. _____
    c. _____
11. a. _____
    b. _____
    c. _____
12. a. _____
    b. _____
    c. _____

Business Applications 3.6

1. Dr. Sun Li will be attending a medical seminar in Boston, Massachusetts. She is figuring her gas budget for the 750-mile trip. Her car averages 25 mi/gal. If gas prices average 133.9¢ a gallon, how much will the gas for this trip cost?

2. Heath Chip Company has been asked to deliver 100 cases of chips to Louise's Catering Service. Each case sells for $5.95. What is the total amount of the invoice presented to the manager at the catering service if 0.06 of the bill is added for sales tax?

3. In preparation for the Kansas Bar Association's Annual Conference, Mr. Cable called the Palace Inn to reserve 750 rooms for three nights. If each room rents for $97.50 a day, what will be the total bill for the conference?

Student Success Hints

READ—Skim for general meaning, get the big picture and become familiar with material (Do you already know anything about the material?). Read for the answers to your questions—highlight, underline or outline/take notes, as you read. In mathematics you must apply what you read, do the examples, and substitute your own numbers.

Activities 3.6

Graph Activity

If the population in the U.S. increases to 300 million, how many gallons of ice cream will be eaten? Use the shortcut method to find the answer.

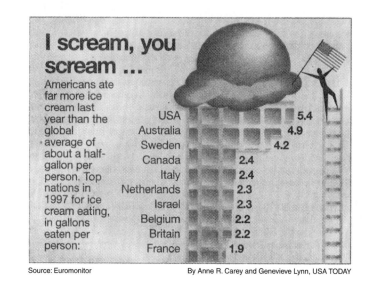

I scream, you scream ...

Americans ate far more ice cream last year than the global average of about a half-gallon per person. Top nations in 1997 for ice cream eating, in gallons eaten per person:

USA	5.4
Australia	4.9
Sweden	4.2
Canada	2.4
Italy	2.4
Netherlands	2.3
Israel	2.3
Belgium	2.2
Britain	2.2
France	1.9

Source: Euromonitor

By Anne R. Carey and Genevieve Lynn, USA TODAY

Challenge Activity

Tots Toy Company has developed a child's game that depends on strategy. A cricket is in a well that is 10 feet deep. The object is to get the cricket out of the well in one day. If the wrong strategy is used, a child will not be able to get the cricket out of the well in one day. Jordan can make the cricket climb 3.7 feet each day. But the cricket slips 1.2 feet each night. About how many days will it take Jordan to get the cricket out of the well.

Internet Activity

Ten employees of Gerased Corporation, where you are working in Personnel, are being honored for fifteen years of company service. In addition to recognition at the company's holiday luncheon, they will receive a dozen roses at their homes. You are in charge of ordering them from

http://www.ftd.com/pricechopper.

Using a shortcut method provide an estimate for sending the flowers.

SKILLBUILDER 3.7

Shortcuts in Division of Decimals

Learning Objectives

* Use shortcuts to divide by multiples of 10.
* Use shortcuts to divide by 25 or multiples of 25.

Dividing by Multiples of 10

To divide by 10, move the decimal point one place to the left. (The decimal point in a whole number is always to the right of the last digit.) For example, $47.693 \div 10 = 4.7693$.

To divide by 100, move the decimal point two places to the left. To divide by 1,000, move the decimal point three places to the left. If necessary, add zeros to the left of the dividend.

If the divisor is a multiple of 10, such as 40, 700, or 3,000, first move the decimal point in the dividend as many places to the left as there are zeros in the divisor. Then divide the resulting number by the nonzero digit (in this case, 4, 7, or 3) to find the quotient.

EXAMPLE

Divide: (a) $11.5894 \div 1,000$

        (b) $96.3 \div 3,000$.

SOLUTION

a. $11.5894 \div 1,000 = 0011.5894 \rightarrow 0.0115894$

b. $96.3 \div 1,000 = 0096.3$

     $0.0963 \div 3 = 0.0321$

Self-Check
Divide: (a) $4,657 \div 100$ and, (b) $1,850 \div 40$.

MATH TIP

Dividing a decimal by a power of 10 can usually be done mentally. You may need a pencil and paper to use some of the other shortcuts, but the work will be easier and quicker than the original problem.

**Self-Check
Answer**
a. 46.57
b. 46.25

Problems

Answers

1._____
_____
_____
_____
_____
_____
_____
_____

2._____
_____
_____
_____
_____
_____
_____
_____

3._____
_____
_____
_____
_____
_____
_____
_____

For each problem, find the quotients and the total. Divide as far as possible. Use the shortcut methods described, and show only your answers. Be sure to align the quotients so that the numbers can be added.

1.

$$854 \div 1{,}000 =$$
$$1{,}467 \div 10 =$$
$$42{,}345 \div 100 =$$
$$8{,}640 \div 500 =$$
$$762 \div 40 =$$
$$1{,}000 \div 10 =$$
$$10{,}000 \div 100 =$$
$$\text{Total} =$$

2.

$$19{,}450 \div 4{,}000 =$$
$$58{,}209 \div 3{,}000 =$$
$$7{,}412 \div 800 =$$
$$164 \div 400 =$$
$$882 \div 700 =$$
$$6{,}500 \div 1{,}000 =$$
$$329 \div 10 =$$
$$\text{Total} =$$

3.

$$\$757.42 \div 80 =$$
$$96.844 \div 400 =$$
$$357.64 \div 2{,}000 =$$
$$327.70 \div 500 =$$
$$900.00 \div 10 =$$
$$65.92 \div 4{,}000 =$$
$$947.65 \div 100 =$$
$$\text{Total} =$$

Calculator Tip

When working with numbers that have a large number of places, your calculator may not be able to display all the places. Shortcuts are useful for getting exact answers in such cases.

Dividing by 25 or a Multiple of 25

To divide by 25, first divide by 100. Then multiply the result by 4. This works because dividing by 25 gives a quotient four times larger than dividing by 100.

To divide by 50, first divide by 100. Then multiply the result by 2.

To divide by 125, first divide by 1,000. Then multiply the result by 8.

To divide by 250, first divide by 1,000. Then multiply the result by 4. To divide by 500, first divide by 1,000. Then multiply the result by 2.

EXAMPLE

Divide: (a) 429 ÷ 25 and (b) $6,160 ÷ 125.

SOLUTION

a. 429 ÷ 100 = 429.

4.29 × 4 = 17.16

b. $6,160 ÷ 1,000 = $6160.

$6.16 × 8 = $49.28

Self-Check
Divide: 357 ÷ 50.

Self-Check Answer
7.14

Problems

Find the quotient in each of the following problems. Use the shortcut methods, and show your work. For money amounts, round your final answers to the nearest cent where necessary.

4. 3,260 ÷ 25

5. $1,729.50 ÷ 50

6. 8,475 ÷ 250

7. $6,750 ÷ 125

8. $418.90 ÷ 125

9. $202.50 ÷ 25

10. 4,650 ÷ 125

11. 217.6 ÷ 250

12. 9,850 ÷ 500

13. 13,500 ÷ 25

14. 33,750 ÷ 500

15. 847.5 ÷ 50

16. $17,295 ÷ 250

17. $21,760 ÷ 125

18. 4,930 ÷ 500

Answers

4._____

5._____

6._____

7._____

8._____

9._____

10._____

11._____

12._____

13._____

14._____

15._____

16._____

17._____

18._____

Business Applications 3.7

Answers

1. _____

2. _____

3. _____

1. The State Department of Transportation needs to decide which state vehicle to send on long trips. Each employee has been asked to report the miles per gallon for his or her issued car. Robert Parker made a 395.5-mi trip on 10 gallons of gas. How many miles per gallon should Robert report for his car?

2. Maria Morian, attorney-at-law, rendered services at a total cost of $1,875 to a client in bankruptcy court. Her hourly fee for such services is $125.00. The bill sent to her client stated the total number of hours worked on the case. For how many hours was the client billed?

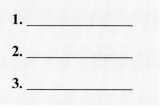

Student Success Hints

Recite—Restate the objective, see if you can answer it. Explain the material to a friend. Complete the self-check problems.

3. Brandon Brittain is restocking the shelves in his baseball card shop. He purchased 375 boxes of baseball cards. If each crate holds 25 boxes, how many crates did he purchase for his shop?

Name:_____ Date:_____

Activities 3.7

Graph Activity

How many residential in-ground and above-ground swimming pools per hundred population are there in Florida? Use the shortcut method and round to the nearest tenth. _____

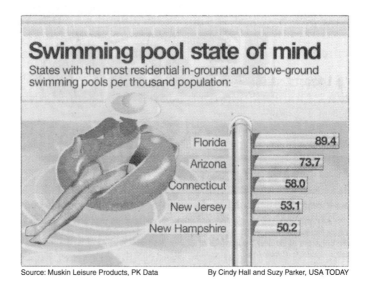

Swimming pool state of mind
States with the most residential in-ground and above-ground swimming pools per thousand population:

Florida	89.4
Arizona	73.7
Connecticut	58.0
New Jersey	53.1
New Hampshire	50.2

Source: Muskin Leisure Products, PK Data   By Cindy Hall and Suzy Parker, USA TODAY

Challenge Activity

Sandra started a savings account at State Bank of New Jersey. She deposited $100 on her birthday and plans to double the deposit every year. How many years will it take Sandra to accumulate $1,500 in her savings account, not counting interest?

Internet Activity

Pretend that it's time to stock up on paper for the office machines. If you were to order Microprint Multi-System paper at

http://www.cmpsatellite.com/cmpsoho.html,

what is the price per sheet for $8\frac{1}{2} \times 11$, 20#, if they are packaged by 500 sheets? What is the price difference, per sheet, to go with the heavier Microprint Desktop stock, instead?

UNIT 4
WORKING WITH FRACTIONS

In this unit, you will study the following Skillbuilders:

Fractions are often used when discussing data, especially in the stock market. As the stock market goes up and down, you will need to be able to add and subtract fractions in order to keep track of your company's stock prices. Which of the stocks shown pay dividends? How many increased in price?

In this unit we study how to add, subtract, multiply, and divide fractions, as well as learn shortcuts for computing with common fractional parts.

NEW YORK (A9) - - Following are complete yearly nationwide composite prices for stocks listed on the New York Stock Exchange.

-01- YRLY-NYSE for JAN 01

Name	Div	Sales (hds)	365-Day High	Low	Close	Chg.
			- A - A -			
AAR	.48	81619	15	$11\frac{1}{2}$	$14\frac{1}{2}$	$+ 2\frac{3}{4}$
ACE Lt n	.43	349641	36	$25\frac{1}{2}$	$31\frac{1}{8}$	$+ 2\frac{1}{2}$
ACMin	1.10	243334	$12\frac{5}{8}$	$10\frac{3}{4}$	$12\frac{1}{4}$	$+ 1\frac{1}{4}$
ACM Op	.80	55391	$10\frac{1}{4}$	$9\frac{1}{8}$	$9\frac{3}{4}$	$+ \frac{3}{8}$
ACM Sc	1.09	355534	$12\frac{1}{4}$	$10\frac{1}{2}$	$11\frac{7}{8}$	$+ 1\frac{1}{4}$
ACMSp	.96	202298	$10\frac{5}{8}$	$8\frac{3}{4}$	10	$+ 1$
ACMMD n	.12	32838	$15\frac{1}{8}$	14	$14\frac{1}{2}$	$- \frac{1}{2}$
ACM MI	1.08	102899	$12\frac{3}{8}$	$9\frac{5}{8}$	$11\frac{1}{4}$	$+ 1\frac{1}{8}$
ACMMM	.72	54633	$9\frac{3}{8}$	$8\frac{5}{8}$	9	$+ \frac{1}{4}$
ACMMu n	.90	35250	$15\frac{1}{8}$	$12\frac{7}{8}$	$13\frac{1}{2}$	$- 1\frac{1}{2}$
ADT wt		110073	2	$\frac{3}{4}$	$\frac{13}{16}$	$- \frac{5}{16}$
ADT		610714	$10\frac{3}{8}$	$6\frac{7}{8}$	$8\frac{7}{8}$	$+ 1\frac{1}{8}$
AFLAC s	.40	405255	34	$24\frac{3}{4}$	$28\frac{1}{2}$	$+ \frac{7}{8}$
AL Lab	.18	183508	$29\frac{3}{8}$	$12\frac{3}{4}$	14	$-11\frac{7}{8}$
AMR		1570478	$72\frac{7}{8}$	$55\frac{1}{2}$	67	$- \frac{1}{2}$
ANR pf	2.67	217	$27\frac{7}{8}$	$25\frac{3}{4}$	$25\frac{3}{4}$	$- \frac{7}{8}$
ANR pfB	2.12	752	27	25	$25\frac{1}{4}$	$+ \frac{1}{8}$
ARCOCh	2.50	74079	$47\frac{1}{4}$	$39\frac{1}{4}$	$43\frac{1}{4}$	$- \frac{1}{2}$
ARX		62267	$3\frac{7}{8}$	$1\frac{3}{4}$	$3\frac{1}{2}$	$+ 1\frac{1}{2}$

MATH CONNECTIONS

Financial Planner

Financial planners often invest money for their clients, helping them to plan for retirement. Most financial planners will invest in either individual stocks or mutual funds. However, to ensure that a client does not sustain a devastating loss, planners keep track of stocks constantly. The New York Stock Exchange lists each stock and gives important information such as the volume of trading (stocks bought and sold). The last four columns list the highest price, lowest price, closing price, and the change from the prior day's closing price. Financial planners will watch for a trend and buy or sell when appropriate. It is interesting to note that the change is in fractions, not decimals.

Math Application

Three years ago, you bought 15 shares of Company ABC, at a price of $\$28\frac{1}{2}$ per share. Your financial planner calls this morning to let you know that your stock's closing price yesterday was $\$29\frac{1}{8}$. If you could sell your stock for $\$29\frac{1}{8}$ per share, how much would you gain? There are actually two ways to compute the gain.

1. Compute the total value of the stock at each price and subtract.

   $(15)(29\frac{1}{8}) - (15)(28\frac{1}{2}) = \$\text{_____}$

2. Find the gain on each share and then multiply by the number of shares to get the total gain.

   $(29\frac{1}{8} - 28\frac{1}{2})(15) = \$\text{_____}$

Critical Thinking Problem

Some stocks require that a fee be paid up front in order to make a purchase. These stocks may be more stable or may outperform other stocks over time. What issues might be pertinent to your decision about which stocks to buy?

Name:_____ Date:_____

Renaming Fractions

Learning Objectives

* **Rename fractions in lower or higher terms.**
* **Rename improper fractions as whole or mixed numbers and vice versa.**
* **Rename fractions as decimals and vice versa.**

Renaming Fractions in Lower or Higher Terms

A fraction is a number written in this form: $\frac{2}{3}$. The number above the line is the **numerator;** the number below the line is the **denominator.** The numerator and denominator are the **terms** of the fraction. A fraction whose numerator is smaller than its denominator is a **proper fraction** and always has a value of less than 1.

To rename a fraction in lower terms, find a number that is contained an even number of times in both the numerator and the denominator. Dividing both the numerator and the denominator by this number will result in a fraction equal in value to the original fraction, but made up of smaller numbers. When no number other than 1 will divide evenly into both terms, a fraction is said to be **in lowest terms.** Find the greatest number that will divide both the numerator and denominator evenly. Then divide both the numerator and denominator by that number.

When a fraction must be renamed as an equivalent fraction with a larger denominator, divide the new denominator (the larger term) by the denominator of the fraction being renamed. Use this quotient to multiply both the numerator and denominator of the fraction to be renamed.

EXAMPLE

a. Write $\frac{12}{28}$ in lowest terms.

b. Rename $\frac{2}{15}$ with a denominator of 60.

SOLUTION

a. $\dfrac{12}{28} = \dfrac{12 \div 4}{28 \div 4} = \dfrac{3}{7}$

Note that no number other than 1 divides both 3 and 7 evenly, so $\frac{3}{7}$ is in lowest terms.

b. $60 \div 15 = 4$

$$\frac{2}{15} = \frac{2 \times 4}{15 \times 4} = \frac{8}{60}$$

Because $\frac{4}{4} = 1$, multiplying or dividing any fraction by $\frac{4}{4}$ does not change the real value of the fraction.

Self-Check Answers

$\frac{3}{7}$

$\frac{15}{20}$

Self-Check

a. Write $\frac{15}{35}$ in lowest terms.

b. Rename $\frac{3}{4}$ with a denominator of 20.

Name:_____ Date:_____

Problems

Rename the following fractions in lowest terms.

1. $\frac{6}{16}$ 2. $\frac{15}{35}$ 3. $\frac{16}{24}$

4. $\frac{27}{81}$ 5. $\frac{18}{72}$ 6. $\frac{9}{45}$

7. $\frac{14}{20}$ 8. $\frac{36}{42}$ 9. $\frac{27}{33}$

10. $\frac{18}{27}$ 11. $\frac{14}{84}$ 12. $\frac{21}{42}$

13. $\frac{14}{42}$ 14. $\frac{26}{39}$ 15. $\frac{15}{25}$

16. $\frac{28}{40}$ 17. $\frac{17}{51}$ 18. $\frac{28}{36}$

Rename the following fractions with the indicated denominator.

19. $\frac{2}{3}$, 27 20. $\frac{3}{8}$, 32 21. $\frac{2}{5}$, 40 22. $\frac{11}{12}$, 36

23. $\frac{4}{32}$, 96 24. $\frac{5}{18}$, 72 25. $\frac{3}{7}$, 56 26. $\frac{4}{15}$, 90

Answers

1.____ 2.____ 3.____ 4.____ 5.____ 6.____ 7.____ 8.____ 9.____ 10.____ 11.____ 12.____ 13.____ 14.____ 15.____ 16.____ 17.____ 18.____ 19.____ 20.____ 21.____ 22.____ 23.____ 24.____ 25.____ 26.____

© Glencoe/McGraw-Hill

Skillbuilder 4.1   Renaming Fractions   161

Renaming Fractions as Mixed Numbers and Vice Versa

A fraction whose numerator is equal to or greater than its denominator is an **improper fraction**. A **mixed number** is a combination of a whole number and a fraction. To rename an improper fraction as a mixed number, divide the denominator into the numerator. The quotient is the whole-number part. The fractional part is formed by the remainder (which becomes the numerator) and the denominator of the original fraction. The fraction should be renamed in lowest terms.

To rename a whole number as a fraction, write the whole number as the numerator with a denominator of 1. To rename a mixed number as an improper fraction, multiply the whole number by the denominator in the fraction, and add the numerator of the fraction to this product. Place this new numerator over the denominator of the mixed number fraction.

EXAMPLE

a. Rename $\frac{30}{4}$ as a mixed number.

b. Rename 15 as an improper fraction.

c. Rename $7\frac{2}{5}$ as an improper fraction.

SOLUTION

a. $\dfrac{30}{4} = 7\frac{2}{4} = 7\frac{1}{2}$

b. $15 = \dfrac{15}{1}$

c. $7\frac{2}{5}$ $\qquad 7 \times 5 + 2 = 37$

$\qquad\qquad\qquad 7\frac{2}{5} = \dfrac{37}{5}$

Self-Check
a. Rename $\frac{29}{4}$ as a mixed number.

b. Rename $8\frac{5}{7}$ as an improper fraction.

Self-Check Answers

$7\frac{1}{4}$

$\frac{61}{7}$

Problems

Rename the following improper fractions as whole or mixed numbers.

27. $\dfrac{12}{10} =$ $\qquad$ **28.** $\dfrac{9}{9} =$ $\qquad$ **29.** $\dfrac{58}{26} =$

30. $\dfrac{117}{20} =$ $\qquad$ **31.** $\dfrac{51}{12} =$ $\qquad$ **32.** $\dfrac{46}{14} =$

Answers

27. _____
28. _____
29. _____
30. _____
31. _____
32. _____

Name:_____ Date:_____

Rename these whole numbers as improper fractions.

33. 75   **34.** 13   **35.** 4   **36.** 166   **37.** 1

Rename the following mixed numbers as improper fractions.

38. $5\frac{4}{6}$ **39.** $16\frac{2}{3}$ **40.** $3\frac{5}{9}$

41. $3\frac{5}{7}$ **42.** $4\frac{5}{11}$ **43.** $12\frac{3}{5}$

Answers

33._____

34._____

35._____

36._____

37._____

38._____

39._____

40._____

41._____

42._____

43._____

Renaming Fractions as Decimals and Vice Versa

To rename a fraction as a proper fraction, write the fraction as the numerator (without a decimal point) over a denominator of 1, plus as many zeros as there are places to the right of the decimal point.

To rename a proper fraction as a fraction, divide the numerator by the denominator. When the division ends in a run-on number, either round the run-on number or express it as a common fraction.

EXAMPLE

Rename 0.56 and $0.5\frac{1}{2}$ as proper fractions in lowest terms.

SOLUTION

$$0.56 = \frac{56}{100} = \frac{56 \div 4}{100 \div 4} = \frac{14}{25}$$

$$0.5\frac{1}{2} = 0.55 = \frac{55}{100} = \frac{55 \div 5}{100 \div 5} = \frac{11}{20}$$

EXAMPLE

Rename $\frac{3}{8}$ and $\frac{1}{6}$ as decimals.

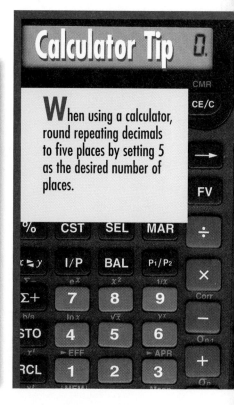

Calculator Tip

When using a calculator, round repeating decimals to five places by setting 5 as the desired number of places.

SOLUTION

$$8\overline{)3.000} = 0.375 \qquad \frac{3}{8} = 0.375$$

$$6\overline{)1.0000} = 0.1666 \qquad \frac{1}{6} = 0.16\frac{2}{3}, \text{ or } 0.167$$

Self-Check

a. Rename $\frac{5}{12}$ as a decimal.

b. Rename 0.78 as a proper fraction in lowest terms.

**Self-Check
Answers**

$0.41\frac{2}{3}$, or 0.417

$\frac{39}{50}$

Problems

Rename the following fractions as decimals. Round to the third decimal place if necessary.

44. $\frac{5}{16}$ **45.** $\frac{4}{32}$ **46.** $\frac{1}{9}$ **47.** $\frac{28}{64}$

Rename the following decimals as proper fractions. Rename the answers in lowest terms.

48. 0.48 **49.** 0.625 **50.** 0.375

51. $0.16\frac{1}{4}$ **52.** $0.22\frac{2}{5}$ **53.** 0.1875

54. 2.125 **55.** $3.366\frac{2}{3}$ **56.** $0.2727\frac{3}{11}$

Answers

44. _____
45. _____
46. _____
47. _____
48. _____
49. _____
50. _____
51. _____
52. _____
53. _____
54. _____
55. _____
56. _____

Business Applications 4.1

1. The president of EDA Brokerage keeps careful track of the stock market. The following stock market summary lists stock measures in decimal form. Change each to a mixed number in lowest terms.

DOW JONES AVERAGES		
20 Utilities	341.49	_____
35 Transportation	1,425.32	_____
70 Stocks	825.21	_____
20 Bonds	94.18	_____
10 Public utility bonds	83.85	_____
20 Industrial bonds	80.03	_____
40 Industrials	2,543.07	_____
Commodity futures	135.69	_____

Answers

2. _____

_____

_____

_____

2. Bikes and More sold 500 bikes in June. Of these, 225 were Spirits, 178 were Hill-Climbers, 75 were Easy Riders, and 22 were special-order racing bikes. Find the fraction (in lowest terms) of each type of bike sold.

3. Chambers Corp. invested in several stocks. On a recent day, they listed the quoted prices of the stocks, which are given in dollars and fractions of a dollar. Write the stock prices in dollars and cents. Round to the nearest cent if necessary.

Stock	Price
PepsiCo	$45\frac{1}{4}$
Wal-Mart	$52\frac{1}{2}$
IBM	$22\frac{7}{8}$
Ford	$26\frac{3}{8}$
Colgate/Palmolive	$39\frac{1}{2}$

4. Several college baseball teams included four batters whose averages for at-bats when they were behind in the count were .329, .303, .298, and .296. Write each batting average as a fraction, reducing to lowest terms.

5. The graph below shows what fields produce most of today's leaders.
Write each fraction as a decimal.

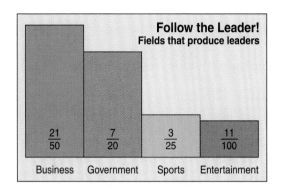

Follow the Leader!
Fields that produce leaders

$\frac{21}{50}$ $\frac{7}{20}$ $\frac{3}{25}$ $\frac{11}{100}$

Business   Government   Sports   Entertainment

6. Fox Research Firm surveyed 520 people 50 and older. They found that
228 people were comfortable recording on VCRs, 136 people were comfortable using ATMs, and 156 were comfortable using computers. Write
a fraction for each type of technology used. Then change the fraction
to a decimal. Round to the nearest hundredth.

Activities 4.1

Graph Activity

Write each decimal as a fraction, reducing to lowest terms.

.250 _____

.321 _____

.335 _____

Rough starts

This year's two expansion teams, Arizona (.373, 47-79) and Tampa Bay (.387, 48-76), are not on pace to have the worst record by an expansion team.

Note: Through Wednesday

.250
.321
.321
.335

N.Y. Mets 1962 (40-120)

Padres 1969 (52-110)

Expos 1969 (52-110)

Blue Jays 1977 (54-107)

Source: Arizona Diamondbacks

By Scott Boeck and Gary Visgaitis, USA TODAY

Challenge Activity

Keep a schedule for one week of how much time you spend each day on your daily activities. Total the hours spent that week on each activity. Then write a fraction representing what part of that week was spent on each activity. Change the fractions to decimals. Were you surprised at any of the results? Determine from this if you need to spend more or less time on each activity.

Internet Activity

Rename $\frac{2}{3}$ with a denominator of 12. Draw a block picture, like a window, to show this concept; fill in panes equaling the numerator. (Hint: Check out **http://forum.swarthmore.edu/paths/fractions/m.fraclessons.html#equiv**) Alter your picture to show the renaming of $\frac{1}{3}$ with a denominator of 12.

SKILLBUILDER 4.2

Finding the Lowest Common Denominator

Learning Objective

• **Find the lowest common denominator of a group of fractions.**

Finding the Lowest Common Denominator

When two or more fractions are to be added or subtracted, all the fractions must have the same denominator. If the denominators are not the same, the fractions must be renamed as fractions with the same denominator. The smallest denominator which is common to all the fractions—the lowest common denominator (LCD)—should be used. One of these methods can be used to find the LCD.

1. If the smaller denominator of a pair of fractions divides evenly into the larger denominator, the larger denominator is the LCD.

2. Multiply the larger denominator by 2. Check to see if the product is divisible by the smaller denominator. If it is not, multiply the larger denominator by 3, 4, and so on, until you find the smallest product divisible evenly by the smaller denominator.

   If the lowest common denominator is not obvious by the previous methods, use the prime number method.

3. *Prime numbers.* A **prime number** is one that is divisible only by 1 and itself. The numbers 2, 3, 5, 7, 11, 13, and 17 are examples of prime numbers. To determine the lowest common denominator by this method, do the following steps.

   a. List all the denominators.

   b. Beginning with the lowest prime number other than 1 as the divisor, divide the denominators and bring down the results.

   c. Continue dividing the results by prime numbers until all the results are reduced to 1.

   d. Find the product of all the prime numbers used. The product is the lowest common denominator.

Problems

Find the lowest common denominator for the following groups of fractions.

Answers

1. _____
2. _____
3. _____
4. _____
5. _____
6. _____
7. _____
8. _____
9. _____
10. _____
11. _____
12. _____

1. $\dfrac{1}{2}, \dfrac{2}{3}, \dfrac{3}{4}, \dfrac{1}{6}$

2. $\dfrac{1}{2}, \dfrac{5}{8}, \dfrac{7}{18}, \dfrac{3}{4}$

3. $\dfrac{2}{3}, \dfrac{1}{4}, \dfrac{4}{15}, \dfrac{7}{8}$

4. $\dfrac{1}{9}, \dfrac{3}{8}, \dfrac{5}{12}, \dfrac{1}{4}$

5. $\dfrac{3}{4}, \dfrac{3}{16}, \dfrac{2}{3}, \dfrac{5}{8}$

6. $\dfrac{3}{5}, \dfrac{4}{15}, \dfrac{11}{12}, \dfrac{2}{3}$

7. $\dfrac{1}{3}, \dfrac{5}{21}, \dfrac{4}{9}, \dfrac{1}{15}$

8. $\dfrac{7}{12}, \dfrac{2}{3}, \dfrac{1}{8}, \dfrac{3}{5}$

9. $\dfrac{2}{3}, \dfrac{1}{15}, \dfrac{3}{4}, \dfrac{5}{8}, \dfrac{1}{2}$

10. $\dfrac{1}{2}, \dfrac{5}{6}, \dfrac{1}{7}, \dfrac{1}{4}, \dfrac{1}{3}$

11. $\dfrac{5}{12}, \dfrac{1}{3}, \dfrac{1}{2}, \dfrac{5}{9}, \dfrac{1}{7}$

12. $\dfrac{1}{6}, \dfrac{7}{30}, \dfrac{3}{16}, \dfrac{5}{32}, \dfrac{3}{8}$

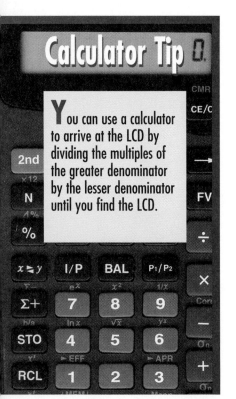

Calculator Tip

You can use a calculator to arrive at the LCD by dividing the multiples of the greater denominator by the lesser denominator until you find the LCD.

170   Unit 4   Working With Fractions

© Glencoe/McGraw-Hill

Business Applications 4.2

Answers

1._____

2._____

1. The Brownville Cab Company recorded the hours worked by each employee on Saturday. Find the LCD of the fractions.

Employee	Hours
Kathy Munoz	$7\frac{1}{2}$
Sun Lee	8
Sean O'Connor	$7\frac{3}{4}$
Rob Thompkins	$6\frac{5}{8}$
Karen Mikovec	$5\frac{21}{26}$

2. Frank Jones, attorney-at-law, kept track of the hours he spent on a divorce case. Find the LCD of the fractions.

Days	Hours
Monday	$2\frac{5}{16}$
Tuesday	$4\frac{2}{3}$
Wednesday	0
Thursday	$1\frac{1}{2}$
Friday	$3\frac{5}{12}$

3. Javiar Lopez is a mechanic at the Tastee Ice Cream Company Shops. In one week he worked the following hours. Find the LCD of the fractions.

Days	Hours
Sunday	$8\frac{1}{2}$
Monday	$9\frac{3}{8}$
Tuesday	10
Wednesday	$9\frac{3}{4}$
Thursday	$10\frac{3}{5}$

4. Compute the lowest common denominator of 3.25, 3.50, and 3.75, for the interest rates listed below.

RATE HIKES

Suppose the fed funds rate had been at 3%. Then this year the Fed pushed it up:

Date	New Rate
Feb. 4	3.25%
March 22	3.50%
April 18	3.75%

Activities 4.2

Graph Activity

Write .688 and .752 in fractions, reducing to lowest terms. _____

_____

Teams of the decades

With two seasons left in the 1990s, the 49ers' regular-season winning percentage is on pace to be the best by any NFL team over a decade. Top teams by decade:

The '90s San Francisco 49ers[1]   .758
The '80s San Francisco 49ers   .688
The '70s Dallas Cowboys   .729
The '60s Green Bay Packers   .722
The '50s Cleveland Browns   .746
The '40s Chicago Bears   .757
The '30s Chicago Bears   .752

Note: Excludes the 1920s when just three teams existed
1 – 1990-97

Source: NFL   By Scott Boeck and Gary Visgaitis, USA TODAY

Challenge Activity

Mr. Brett, payroll supervisor has asked you and two classmates to calculate the paychecks for your department. Use your classmates' and/or your hourly rate and number of hours worked as data for this project. Calculate the number of hours worked the past week plus overtime of $2\frac{1}{2}$ hours, $4\frac{1}{4}$ hours, and $3\frac{1}{3}$ hours for three days. Use time and a half as the amount of overtime pay. What is the regular salary? Overtime salary? Total salary? Remember you would subtract taxes and other payroll deductions from the total salary to reach the amount of your check. Calculate your pay and two classmates. What is the total payroll for the department?

Internet Activity

Nullo is a fast-paced game for two to four players, at

http://forum.swarthmore.edu/paths/fractions/nullo.html

```
W       7   5      W       42  20
     I       35         I        2
49       L          4       L
 3   9       D     18  28       D
```

SKILLBUILDER 4.3

Adding Fractions, Mixed Numbers, and Decimal Equivalents

Learning Objectives

- Add fractions or mixed numbers with like or unlike denominators.
- Add fractions or mixed numbers by writing the fraction in decimal form.

Adding Fractions and Mixed Numbers

Fractions can be added only when they have like denominators. The total of the numerators of the addends is the numerator of the sum. The common denominator of the addends becomes the denominator of the sum. Rename the answer in lowest terms if necessary.

When the denominators are not all alike, the lowest common denominator must be found; all fractions must then be renamed in like terms.

EXAMPLE

a. Add: $\frac{7}{24} + \frac{11}{24}$.

b. Add: $\frac{4}{5} + \frac{3}{5}$.

c. Add: $27\frac{3}{8} + 13\frac{5}{16} + 3\frac{1}{3} + 2\frac{5}{6}$.

SOLUTION

a.
$$\frac{7}{24}$$
$$+\frac{11}{24}$$
$$\overline{\frac{18}{24}} = \frac{3}{4}$$

b.
$$\frac{4}{5}$$
$$+\frac{3}{5}$$
$$\overline{\frac{7}{5}} = 1\frac{2}{5}$$

Self-Check Answer

$14\frac{1}{8}$

c. The LCD is 48. Each mixed number is rewritten with a denominator of 48. Then the fractions are added. Finally, the whole numbers are added, and the resulting mixed number is written in lowest terms.

$$27\frac{3}{8} = 27\frac{18}{48}$$
$$13\frac{5}{16} = 13\frac{15}{48}$$
$$3\frac{1}{3} = 3\frac{16}{48}$$
$$2\frac{5}{6} = 2\frac{40}{48}$$
$$\overline{}$$
$$45\frac{89}{48} = 45 + 1\frac{41}{48} = 46\frac{41}{48}$$

Self-Check

Add the following fractions: $6\frac{7}{12} + 2\frac{3}{8} + 5\frac{1}{6}$.

Rename the answer in lowest terms.

Name: _____ Date: _____

Problems

Add the following groups of fractions. Rename the answers in lowest terms.

1. $2\frac{1}{4}$ =
    $4\frac{3}{8}$ =
    $3\frac{11}{32}$ =
    $2\frac{1}{2}$ =
    $+ 3\frac{1}{16}$ =

2. $7\frac{7}{18}$ =
    $3\frac{5}{6}$ =
    $3\frac{1}{3}$ =
    $5\frac{5}{12}$ =
    $+ 2\frac{5}{9}$ =

3. $16\frac{1}{8}$ =
    $9\frac{2}{5}$ =
    $2\frac{7}{15}$ =
    $3\frac{7}{9}$ =
    $+ 1\frac{1}{16}$ =

4. $4\frac{3}{5}$ =
    $2\frac{3}{7}$ =
    $5\frac{3}{4}$ =
    $2\frac{3}{14}$ =
    $+ 1\frac{1}{8}$ =

5. $4\frac{3}{8}$ =
    $3\frac{2}{3}$ =
    $3\frac{7}{12}$ =
    $2\frac{4}{9}$ =
    $+ 3\frac{1}{2}$ =

6. $5\frac{7}{30}$ =
    $2\frac{1}{11}$ =
    $3\frac{5}{6}$ =
    $2\frac{2}{3}$ =
    $+ 6\frac{2}{5}$ =

7. $4\frac{1}{4}$ =
    $2\frac{1}{3}$ =
    $3\frac{1}{3}$ =
    $9\frac{3}{4}$ =
    $+ 1\frac{1}{8}$ =

8. $12\frac{1}{6}$ =
    $9\frac{2}{3}$ =
    $5\frac{5}{6}$ =
    $11\frac{5}{24}$ =
    $+ 10\frac{1}{8}$ =

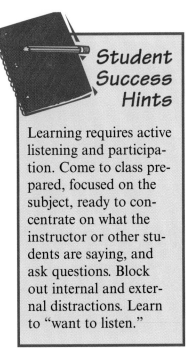

Student Success Hints

Learning requires active listening and participation. Come to class prepared, focused on the subject, ready to concentrate on what the instructor or other students are saying, and ask questions. Block out internal and external distractions. Learn to "want to listen."

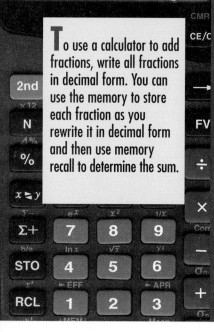

Calculator Tip

To use a calculator to add fractions, write all fractions in decimal form. You can use the memory to store each fraction as you rewrite it in decimal form and then use memory recall to determine the sum.

Adding Fractions Using Decimal Equivalents

Fractions can also be added by using their decimal equivalents. A fraction is expressed as its decimal equivalent by dividing the denominator into the numerator.

Sometimes a group of numbers contains both decimal fractions and common fractions. When this occurs, express all the numbers either as decimal fractions or as common fractions before proceeding with the addition.

EXAMPLE

a. Add in decimal form and fraction form: $6\frac{1}{2} + 3\frac{5}{8} + 4\frac{1}{16} + 2\frac{3}{4}$.

b. Add: $0.37\frac{1}{2} + 0.42\frac{1}{4} + \frac{5}{8} + 0.48\frac{2}{3}$. Round to four decimal places, where necessary.

SOLUTION

a. Add as Decimals

$$6.5000$$
$$3.6250$$
$$4.0625$$
$$\underline{+\ 2.7500}$$
$$16.9375$$

Add as Fractions

$$6\frac{1}{2} = 6\frac{8}{16}$$
$$3\frac{5}{8} = 3\frac{10}{16}$$
$$4\frac{1}{16} = 4\frac{1}{16}$$
$$\underline{+\ 2\frac{3}{4} = 2\frac{12}{16}}$$
$$15\frac{31}{16} = 16\frac{15}{16}$$

Note: 0.9375 is equivalent to $\frac{15}{16}$.

b.
$$0.37\frac{1}{2} = 0.3750$$
$$0.42\frac{1}{4} = 0.4225$$
$$\frac{5}{8} = 0.6250$$
$$\underline{+\ 0.48\frac{2}{3} = 0.4867}$$
$$1.9092$$

Self-Check
Answer
4.49286

Self-Check
Add: $0.62\frac{1}{2}$, $\frac{1}{7}$, and $3.7\frac{1}{4}$. Round to five decimal places, where necessary.

Name:_____ Date:_____

Problems

Add the following groups of numbers by renaming all addends as fractions.
Round to five decimal places, where necessary.

9. $\frac{5}{16} =$

$0.37\frac{1}{2} =$

$0.15\frac{1}{8} =$

$+ \quad \frac{1}{32} =$

10. $7.16\frac{3}{4} =$

$6.25\frac{2}{5} =$

$8.9\frac{5}{6} =$

$+ 6.02\frac{1}{4} =$

11. $0.87\frac{1}{2} =$

$6.7\frac{1}{4} =$

$4.7\frac{1}{2} =$

$+ \quad \frac{5}{8} =$

12. $0.37\frac{3}{4} =$

$5.43\frac{5}{6} =$

$0.93\frac{1}{5} =$

$+ 9\frac{1}{8} =$

13. $\frac{5}{8} =$

$0.6\frac{1}{4} =$

$0.34\frac{3}{8} =$

$+ 45\frac{1}{4} =$

14. $0.45\frac{1}{2} =$

$\frac{1}{16} =$

$3.3\frac{1}{4} =$

$+ 4\frac{3}{16} =$

15. $7.15\frac{3}{4} =$

$0.94 =$

$2.33\frac{1}{16} =$

$+ 4.37\frac{1}{2} =$

16. $0.92\frac{1}{16} =$

$2.03\frac{1}{2} =$

$8.43\frac{3}{32} =$

$+ 10\frac{3}{16} =$

17. $\frac{3}{32} =$

$0.42\frac{1}{2} =$

$3.25\frac{2}{5} =$

$+ 55\frac{1}{4} =$

Answers

9. _____
10. _____
11. _____
12. _____
13. _____
14. _____
15. _____
16. _____
17. _____

Business Applications 4.3

1. During inventory, Week's Paint Store located the following quantities of paint stored in the stockroom: $22\frac{1}{2}$ gal of white, $23\frac{1}{2}$ gal of sky blue, $34\frac{1}{4}$ gal of royal blue, $11\frac{3}{4}$ gal of bright yellow, and $35\frac{1}{4}$ gal of beige. How much paint is stored in the stockroom?

$$22\frac{1}{2} =$$
$$23\frac{1}{2} =$$
$$34\frac{1}{4} =$$
$$11\frac{3}{4} =$$
$$+\ 35\frac{1}{4} =$$

2. Luanda Jackson, a plumber, worked $2\frac{1}{4}$ hr of overtime on Tuesday, $1\frac{1}{2}$ hr on Wednesday, $2\frac{2}{3}$ hr on Friday, and $7\frac{3}{4}$ hr on Saturday. Find the total number of overtime hours she worked.

$$2\frac{1}{4} =$$
$$1\frac{1}{2} =$$
$$2\frac{2}{3} =$$
$$+\ 7\frac{3}{4} =$$

3. Nan's fruit and nut store sold $8\frac{5}{8}$ pounds of pistachios, $19\frac{2}{5}$ pounds of peanuts, $6\frac{3}{10}$ pounds of cashews, $23\frac{1}{2}$ pounds of pecans, and $10\frac{1}{4}$ pounds of mixed nuts in one day. How many pounds of nuts did Nan sell?

$$8\frac{5}{8} =$$
$$19\frac{2}{5} =$$
$$6\frac{3}{10} =$$
$$23\frac{1}{2} =$$
$$+\ 10\frac{1}{4} =$$

Activities 4.3

Graph Activity

Look at the graph below. 1. Change these decimals to mixed numbers with the same denominator; 2. Add the mixed numbers and reduce to lowest terms; 3. Convert the answer in problem 2 back to a decimal; 4. What was the total number of weekly listeners for these 6 radio personalities?

Answers

1._____

_____

_____

_____

_____

_____

2._____

3._____

4._____

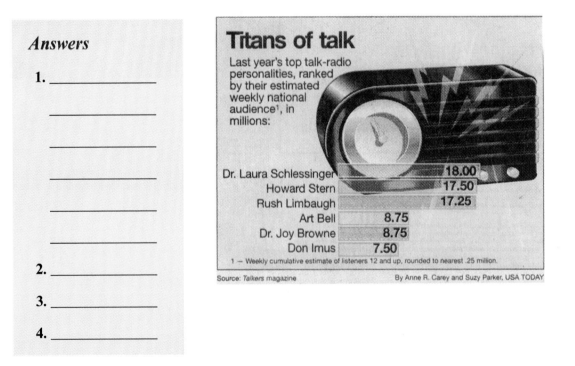

Titans of talk

Last year's top talk-radio personalities, ranked by their estimated weekly national audience[1], in millions:

Dr. Laura Schlessinger	18.00
Howard Stern	17.50
Rush Limbaugh	17.25
Art Bell	8.75
Dr. Joy Browne	8.75
Don Imus	7.50

1 — Weekly cumulative estimate of listeners 12 and up, rounded to nearest .25 million.

Source: *Talkers* magazine   By Anne R. Carey and Suzy Parker, USA TODAY

Challenge Activity

At the grocery store, look for different items that have decimals or fractions in their weights. List the item and its decimal weight. Once you have listed ten items, change the decimals to fractions. Is the decimal or fraction easier for you to understand the size? Do some products use decimals more often than others? Did you find any items that used fractions instead of decimals? What kind of items used fractions? Why do you suppose that some products use fractions and others use decimals?

Internet Activity

When adding fractions, you can check your answer with your calculator. Find out how at

http://www.math.utep.edu/sosmath/ algebra/fraction/frac3/frac37/ frac37.html

SKILLBUILDER 4.4

Subtracting Fractions and Mixed Numbers

Learning Objective

- **Find the difference of two fractions or mixed numbers.**

Subtracting Fractions and Mixed Numbers

As in addition, like denominators must be used in order to subtract fractions. Subtract the numerator of the subtrahend from the numerator of the minuend. The common denominator of the two fractions will be the denominator of the difference. Rename in lowest terms if necessary.

When mixed numbers are subtracted, the fraction in the subtrahend may be larger than the fraction in the minuend. If so, subtract one unit from the whole number in the minuend. Change it to a fraction and add it to the fraction in the minuend. Once the borrowing is accomplished, subtract the whole numbers and the numerators of the fractions.

MATH TIP

To reduce fractions, use the rules of divisibility. These include the following: A number is divisible by 2 if it is an even number, 3 if the sum of the digits is divisible by 3, and 5 if the number ends in a 0 or a 5.

a. Subtract: $\frac{7}{8} - \frac{3}{4}$.

b. Subtract: $9\frac{3}{8} - 5\frac{5}{8}$.

SOLUTION

a.
$$\frac{7}{8} = \frac{7}{8}$$
$$-\frac{3}{4} = \frac{6}{8}$$
$$\overline{\qquad \quad \frac{1}{8}}$$

b. Since $\frac{5}{8}$ is greater than $\frac{3}{8}$, we must rewrite one unit from the whole number, 9, as a fraction and add it to $\frac{3}{8}$.

$$9\frac{3}{8} = 8 + 1 + \frac{3}{8} = 8 + \frac{8}{8} + \frac{3}{8} = 8\frac{11}{8}$$
$$-5\frac{5}{8} = 5\frac{5}{8} \qquad\qquad = 5\frac{5}{8} \qquad\qquad = 5\frac{5}{8}$$
$$\overline{\qquad\qquad\qquad\qquad\qquad\qquad\qquad\qquad 3\frac{6}{8}, \text{ or } 3\frac{3}{4}}$$

**Self-Check
Answer**

$5\frac{43}{48}$

Self-Check

Subtract: $18\frac{5}{6} - 12\frac{15}{16}$.

Problems

Find the difference in each problem. Rename in lowest terms, if necessary.

1. $24\frac{3}{8} =$
   $-11\frac{5}{6} =$

2. $35\frac{1}{12} =$
   $-13\frac{5}{12} =$

3. $7\frac{2}{9} =$
   $-3\frac{5}{16} =$

4. $54\frac{2}{5} =$
   $-39\frac{7}{10} =$

5. $36\frac{6}{35} =$
   $-24\frac{4}{5} =$

6. $5\frac{11}{22} =$
   $-2\frac{7}{12} =$

7. $49\frac{1}{2} =$
   $-29\frac{19}{36} =$

8. $24\frac{7}{27} =$
   $-10\frac{5}{6} =$

9. $9\frac{1}{2} =$
   $-7\frac{17}{18} =$

10. $8\frac{1}{2} =$
    $-6\frac{5}{8} =$

11. $68\frac{5}{8} =$
    $-9\frac{15}{16} =$

12. $44\frac{3}{16} =$
    $-14\frac{19}{48} =$

13. $29\frac{2}{3} =$
    $-13\frac{1}{4} =$

14. $68\frac{5}{8} =$
    $-36.37\frac{1}{2} =$

15. $88\frac{1}{3} =$
    $-22\frac{1}{4} =$

Answers

1. _____
2. _____
3. _____
4. _____
5. _____
6. _____
7. _____
8. _____
9. _____
10. _____
11. _____
12. _____
13. _____
14. _____
15. _____
16. a. _____
    b. _____

16. **a.** How much must be added to $\frac{1}{4}$ to get $\frac{1}{2}$?

    b. How much must be added to $\frac{3}{8}$ to get $\frac{7}{8}$?

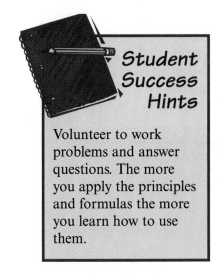

Student Success Hints

Volunteer to work problems and answer questions. The more you apply the principles and formulas the more you learn how to use them.

Answers

17. _____

18. _____

19. _____

20. _____

21. _____

22. _____

23. _____

24. _____

25. _____

26. _____

27. _____

28. _____

17. $222\frac{8}{15} =$
$-\ 173\frac{21}{25} =$

18. $10\frac{2}{3} =$
$-\ 5\frac{5}{6} =$

19. $25\frac{17}{24} =$
$-\ 10\frac{31}{36} =$

20. $26\frac{7}{20} =$
$-\ 24\frac{11}{14} =$

21. $65\frac{8}{35} =$
$-\ 16\frac{11}{14} =$

22. $267\frac{4}{7} =$
$-\ 203\frac{2}{9} =$

23. $82\ \ =$
$-\ 9\frac{5}{22} =$

24. $173\frac{21}{25} =$
$-\ 94\frac{8}{15} =$

Calculator Tip

T o subtract mixed numbers using a calculator, write each number as a decimal by writing the fraction as a decimal and then adding the whole number.

25. $\frac{9}{19} =$
$-\ \frac{3}{17} =$

26. $365\frac{27}{50} =$
$-\ 123\frac{39}{40} =$

27. $47\ \ =$
$-\ 8\frac{3}{27} =$

28. $\frac{11}{15} =$
$-\ \frac{7}{10} =$

Business Applications 4.4

1. Dr. White has a patient who needs to reach his desired weight of 182 lb. To reach this weight, the patient needs to lose 15 lb. If the patient loses $5\frac{1}{2}$ lb the first week and $3\frac{1}{4}$ lb the second week, how much more weight must he lose to reach his desired weight?

$$5\frac{1}{2} = \qquad\qquad 15 =$$
$$+\ 3\frac{1}{4} = \qquad\qquad -\ 8\frac{3}{4} =$$

2. Hutchens Recycling Company collects steel to be recycled. To determine how much steel the company collected, subtract the amount in the Weight of Cart column from the gross weight to find the weight of the load in each cart. Then find the total weight of the five loads.

	Cart No.	Gross Weight	Weight of Cart	Weight of Load
	HUTCHENS RECYCLING COMPANY			
a.	15	$3,856\frac{1}{2}$ lb	$1,609\frac{7}{8}$ lb	_____
b.	22	$4,516$ lb	$1,637\frac{3}{4}$ lb	_____
c.	26	$4,763\frac{1}{8}$ lb	$1,609\frac{7}{8}$ lb	_____
d.	31	$4,873\frac{1}{4}$ lb	$1,609\frac{7}{8}$ lb	_____
e.	35	$4,529\frac{7}{8}$ lb	$1,637\frac{3}{4}$ lb	_____
			Total weight	_____

3. Gem Industries' stock closed on Friday at a price of $42\frac{5}{8}$. On Monday, this stock closed at a price of $39\frac{7}{8}$. By how much did the market value of the stock drop on Monday?

4. Raymond's Fish Hatchery made a shipment to the locker plant. To determine the weight of fish sent, subtract the bucket weight from the gross weight of each catch to find the weight of the fish. Then find the total of the amounts in the Weight of Fish column.

RAYMOND'S FISH HATCHERY

	Gross Weight	Bucket Weight	Weight of Fish
a.	$327\frac{5}{16}$ lb	$42\frac{1}{2}$ lb	_____
b.	$321\frac{5}{8}$ lb	$42\frac{1}{2}$ lb	_____
c.	$337\frac{7}{8}$ lb	$45\frac{3}{4}$ lb	_____
d.	$344\frac{1}{4}$ lb	$45\frac{3}{4}$ lb	_____
e.	$323\frac{3}{8}$ lb	$42\frac{1}{2}$ lb	_____
		Total weight	_____

Name:_____ Date:_____

5. Marcia is building a frame that needs to be $12\frac{1}{2}$ inches in length. The wood she is using is $14\frac{1}{3}$ inches in length. How many inches does Marcia need to cut the wood to make it $12\frac{1}{2}$ inches in length?

Answers

5. _____

6. _____

7. _____

8. _____

6. Cyber Company's common stock was selling at $22\frac{1}{4}$ this morning and closed at $23\frac{1}{8}$. By how much did the stock rise from the morning to closing?

7. Mara had 16 ounces of jellybeans. She gave $\frac{1}{2}$ to one friend and $\frac{1}{4}$ to her brother. How many ounces does Mara have left?

8. On Friday, Ron's Paint Company painted $\frac{1}{4}$ of a house. On Monday, the painters completed another $\frac{1}{3}$ of the house. How much of the house is still left to paint?

Activities 4.4

Graph Activity

Sam and Chris Rayos are living with their Aunt Rita while attending college. Sam will pay $\frac{1}{3}$ of the mortgage payment and Chris will pay $\frac{1}{4}$. What fraction of the mortgage payment is left for Aunt Rita to pay? _____

How couples split the bills

How married couples pay their bills:

Split bills by income level **15%**

Split 50-50 regardless of income **15%**

Pay from joint income pool **67%**

Don't know **3%**

Source: Yankelovich Partners for Lutheran Brotherhood   By Cindy Hall and Grant Jerding, USA TODAY

Challenge Activity

Two world records for the men's 1-mile were 4:01.6 and 3:44.39. Convert the race times to seconds, changing the decimals to fractions. Find the difference in times. _____

Internet Activity

Here's a brainteaser from

http://forum.swarthmore.edu/paths/fractions/fracsqrs.html

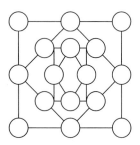

SKILLBUILDER 4.5

Multiplying Fractions and Mixed Numbers

Learning Objectives

- **Find the product of two or more fractions.**
- **Find the product of two or more mixed numbers.**

Multiplying Fractions

To multiply fractions, there is no need to convert denominators to like terms; any combination of factors can be multiplied. Multiply numerator by numerator and denominator by denominator. Rename the resulting product in simplest terms. If an improper fraction results, change it to a mixed or whole number.

If **cross factors** (the numerator of one fraction and the denominator of the other fraction) can be evenly divided before multiplying, do so in order to simplify the multiplication. This process of dividing cross factors is called **cancellation.** If either the multiplicand or the multiplier is a whole number, express the whole number with a denominator of 1.

Self-Check Answers

$\frac{3}{8}$

$2\frac{2}{3}$

EXAMPLE

a. Multiply: $\frac{3}{8} \times \frac{2}{9}$.

b. Multiply: $5 \times \frac{2}{15}$.

SOLUTION

a. $\dfrac{\overset{1}{\cancel{3}}}{\underset{4}{\cancel{8}}} \times \dfrac{\overset{1}{\cancel{2}}}{\underset{3}{\cancel{9}}} = \dfrac{1}{12}$

b. $5 \times \dfrac{2}{15} = \dfrac{\overset{1}{\cancel{5}}}{1} \times \dfrac{2}{\underset{3}{\cancel{15}}} = \dfrac{2}{3}$

Self-Check

a. Multiply: $\frac{3}{5} \times \frac{5}{8}$.

b. Multiply: $12 \times \frac{2}{9}$.

Problems

Find the products in the following problems. Rename the answers in simplest terms.

1. $\dfrac{5}{16} \times \dfrac{4}{15} =$

2. $\dfrac{8}{9} \times \dfrac{5}{24} =$

3. $\dfrac{7}{12} \times \dfrac{8}{49} =$

4. $\dfrac{5}{6} \times \dfrac{3}{4} =$

5. $\dfrac{4}{7} \times \dfrac{3}{7} =$

6. $3 \times \dfrac{5}{12} =$

7. $\dfrac{8}{27} \times 9 =$

8. $6 \times \dfrac{5}{18} =$

9. $\dfrac{3}{5} \times \dfrac{2}{3} =$

10. $\dfrac{2}{5} \times \dfrac{3}{10} =$

11. $5 \times \dfrac{3}{25} =$

12. $\dfrac{5}{8} \times \dfrac{9}{16} =$

13. $\dfrac{1}{2} \times \dfrac{3}{8} =$

14. $7 \times \dfrac{4}{21} =$

15. $\dfrac{11}{25} \times \dfrac{10}{33} =$

16. $\dfrac{8}{21} \times \dfrac{3}{16} \times \dfrac{2}{5} =$

17. $\dfrac{5}{34} \times \dfrac{4}{15} \times \dfrac{7}{8} =$

18. $\dfrac{8}{51} \times \dfrac{3}{16} \times \dfrac{5}{12} =$

Answers

1. _____
2. _____
3. _____
4. _____
5. _____
6. _____
7. _____
8. _____
9. _____
10. _____
11. _____
12. _____
13. _____
14. _____
15. _____
16. _____
17. _____
18. _____

Multiplying Mixed Numbers

To multiply mixed numbers, rename the factors as improper fractions. Use cancellation if possible, and rename the answer in lowest terms.

Decimal fractions may be multiplied by common fractions and mixed numbers.

EXAMPLE

a. Multiply: $3\frac{3}{8} \times 2\frac{4}{9}$.

b. Multiply: $\$24.16 \times 2\frac{1}{4}$.

SOLUTION

a. $3\frac{3}{8} \times 2\frac{4}{9} = \frac{27}{8} \times \frac{22}{9}$

$= \frac{\overset{3}{\cancel{27}}}{\underset{4}{\cancel{8}}} \times \frac{\overset{11}{\cancel{22}}}{\underset{1}{\cancel{9}}} = \frac{3}{4} \times \frac{11}{1} = \frac{33}{4} = 8\frac{1}{4}$

b. $\$24.16 \times 2\frac{1}{4} = \frac{\overset{6.04}{\cancel{\$24.16}}}{1} \times \frac{9}{\underset{1}{\cancel{4}}} = \frac{\$54.36}{1} = \$54.36$

Self-Check Answers
$2\frac{17}{20}$
$\$236.44$

Self-Check

a. Multiply: $2\frac{2}{5} \times 1\frac{3}{16}$.

b. Multiply: $\$51.40 \times 4\frac{3}{5}$.

Problems

Find the product in each of the following problems. Show the details of your work, and check each product.

19. $12\frac{1}{4} \times 2\frac{2}{7} =$

20. $24\frac{1}{4} \times 15\frac{1}{3} =$

21. $36\frac{2}{3} \times 21\frac{3}{8} =$

22. $12\frac{7}{8} \times 6\frac{3}{4} =$

Answers

19. _____

20. _____

21. _____

22. _____

23. $3\frac{1}{3} \times 4\frac{1}{8} =$ **24.** $6\frac{3}{4} \times 1\frac{1}{9} =$

25. $8\frac{1}{6} \times 2\frac{1}{7} =$ **26.** $11\frac{1}{4} \times 1\frac{1}{15} =$

27. $1\frac{1}{2} \times 2\frac{5}{6} =$ **28.** $6\frac{1}{2} \times 3\frac{1}{4} =$

29. $2\frac{1}{3} \times 3\frac{1}{2} =$ **30.** $3\frac{1}{3} \times 4\frac{1}{5} =$

Answers

23._____

24._____

25._____

26._____

27._____

28._____

29._____

30._____

Business Applications 4.5

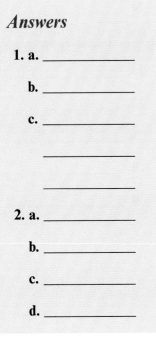

Answers

1. a. _____

   b. _____

   c. _____

   _____

   _____

2. a. _____

   b. _____

   c. _____

   d. _____

1. Derek, Andrea, and Kail formed a partnership to market Worldwide Distributors. They agreed to take a share of the profits each month for personal expenses and to put the balance back into the business. The shares agreed upon were $\frac{2}{5}$, $\frac{1}{4}$, and $\frac{1}{8}$, respectively. The profit for the month of June was $22,520.

   a. What fraction of the profit did their combined shares equal?

   b. What fraction of the profit did they put back into the business?

   c. What was the value of each of their shares?

2. Rabin Market's monthly expenses are $24,000. Of this $24,000, $\frac{1}{3}$ is for salaries, $\frac{1}{16}$ is for utilities, $\frac{1}{8}$ is for advertising, and $\frac{1}{12}$ is for general maintenance. What is the expense each month for the following?

   a. Salaries
   b. Utilities
   c. Advertising
   d. General maintenance

3. Last year Carlos Larson jogged $3\frac{5}{8}$ mi each day. Now he jogs $2\frac{1}{2}$ times this distance. How many miles does he now run each day?

4. Brian Lyn uses his car for business. Last year he drove 21,876 mi, using $1,393\frac{4}{10}$ gal of gasoline. If the average price per gallon was $1.05\frac{9}{10}$ a gallon, how much did he spend on gasoline? (Round your final answer to the nearest cent.)

5. The Hardings are ordering new carpeting costing $24.75 a square yard ($2.75 a square foot). The areas to be carpeted are the living room, $21\frac{1}{2}$ ft long by $18\frac{1}{4}$ ft wide; and the dining room, $18\frac{1}{4}$ ft long by $14\frac{3}{4}$ foot wide. What will be the total cost of carpet for these rooms? (Multiply length times width to get area. Determine the total area first; then work out the cost. Round your answer to the nearest cent.)

6. For each pound of pistachio nuts purchased, Taft's Nuts is offering customers an extra $\frac{1}{4}$ lb free. Alice Tam bought 5 lb of the nuts. How many pounds will she get free?

7. Lucy Ford has been hired by the Carter Corporation to do odd jobs. She will be paid $8.50 an hour. On Monday, she worked $4\frac{1}{2}$ hr. On Tuesday, she worked $6\frac{1}{4}$ hr. On Friday, she worked $3\frac{3}{4}$ hr. How much did she earn for the 3 days that she worked?

8. Widget Industries has rented booth space at the International Trade Show Center. In the past, Widget has spent $\frac{7}{16}$ of every dollar on exhibit construction and space rental, $\frac{1}{8}$ for transportation, $\frac{11}{32}$ on specialty advertising and personnel, and the balance on cleanup and removal of the exhibit after the show. If Widget has budgeted $228,000 for the coming show, how much will it spend on each of these items?

9. The current market values of May Garcia's stocks are listed here. What is the value of her investment in each stock, and what is the total value of her portfolio? (In computing your answers, change any fractional parts of a dollar to cents, rounding off to three decimal places, where necessary.)

   a. Stock A   54 shares at 48\frac{1}{8}$ a share

   b. Stock B   300 shares at 12\frac{5}{8}$ a share

   c. Stock C   80 shares at 39\frac{1}{2}$ a share

   d. Stock D   84 shares at 68\frac{7}{8}$ a share

   e. Stock E   312 shares at 32\frac{3}{4}$ a share

   f. Total

10. Andrew Porter died without leaving a will. Under the laws of his state, $\frac{1}{2}$ of his personal estate goes to his widow and the remainder is divided equally among his eight children. The estate has a value of $284,800.

   a. How much would Porter's widow receive?

   b. How much would each child receive?

Answers

10. a. _____

   b. _____

11. a. _____

   b. _____

12. a. _____

   b. _____

11. The City of Laird has an opportunity to purchase land for a park. The section in question measures $\frac{7}{8}$ mile by $\frac{5}{12}$ mile.

   a. If there are 640 acres to the square mile, how many acres are contained in the plot?

   b. When federal funds were withdrawn, the city had only enough money to buy $77\frac{1}{9}$ acres of the available land. How many acres were they unable to buy?

Calculator Tip

Memory keys allow you to store amounts for problems containing two or more parts.

12. A rule of thumb says that your monthly mortgage payment, including interest, should not exceed $\frac{7}{25}$ of your gross monthly income.

   a. Bruce Angler earns $43,500 a year. How much of a mortgage payment can he afford per month?

   b. Yvette Saunders earns $51,800 a year. How much of a mortgage payment can she afford per month?

Activities 4.5

Graph Activity

If fuel had cost 7.5 cents less per gallon, how much would an airline have saved? (Hint: Each cent decrease saves $2.5 million.) _____

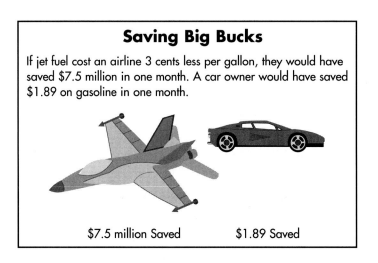

Saving Big Bucks

If jet fuel cost an airline 3 cents less per gallon, they would have saved $7.5 million in one month. A car owner would have saved $1.89 on gasoline in one month.

$7.5 million Saved $1.89 Saved

Challenge Activity

As a first time investor you have $500.00 to invest on two stocks. Look through the stock market section in a newspaper to locate your stocks. Two percent of the $500.00 will be used for the broker's commission. Choose stock with a closing price containing a fraction. It is not necessary to use the entire $500.00. Which stocks did you buy? What was the closing price per share of each stock? How many shares of each stock did you purchase? What was the total price for each stock? (Remember to round to the nearest cent.) What was the total amount spent for both stocks and the broker's commission? Keep this information for future problems.

Use the following table/template for all stock market projects in this unit.

Date	Close	#Purchased	Broker's Commission	Cost of Stock	Total Cost

Internet Activity

Any integer can be written as a fraction

(http://www.math.utep.edu/sosmath/algebra/frac3/frac31/frac318/frac318.html).

SKILLBUILDER 4.6

Dividing Fractions and Mixed Numbers

Learning Objectives

- **Divide one fraction by another.**
- **Divide one mixed number by another.**

Dividing Fractions

To divide common fractions, invert (turn upside down) the divisor and multiply. Rename the quotient in simplest terms. When either the dividend or the divisor is a whole number, write it as that number over 1 before dividing. Do not use cancellation until *after* the divisor has been inverted.

EXAMPLE

Find each quotient.

a. $\dfrac{3}{5} \div \dfrac{9}{16}$

b. $\dfrac{5}{12} \div 3$

MATH TIP

A number is divisible by 6 if the number is divisible by both 2 and 3; by 9 if the sum of the digits is divisible by 9; and by 10 if the number ends in a 0.

SOLUTION

a. $\dfrac{3}{5} \div \dfrac{9}{16} = \dfrac{3}{5} \times \dfrac{\overset{1}{\cancel{3}}}{5} \times \dfrac{16}{\underset{3}{\cancel{9}}} = \dfrac{16}{15} = 1\dfrac{1}{15}$

b. $\dfrac{5}{12} \div 3 = \dfrac{5}{12} \div \dfrac{3}{1} = \dfrac{5}{12} \times \dfrac{1}{3} = \dfrac{5}{36}$

Self-Check

Find the quotient: $4 \div \dfrac{2}{9}$.

Self-Check Answer
18

Problems

Find the quotients in the following problems. Rename the answers in simplest terms if necessary.

Answers

1. _____
2. _____
3. _____
4. _____
5. _____
6. _____
7. _____
8. _____
9. _____
10. _____
11. _____
12. _____
13. _____
14. _____
15. _____
16. _____
17. _____
18. _____

1. $\dfrac{3}{8} \div \dfrac{7}{9} =$

2. $\dfrac{2}{7} \div \dfrac{4}{11} =$

3. $\dfrac{5}{12} \div 15 =$

4. $\dfrac{5}{6} \div \dfrac{5}{12} =$

5. $\dfrac{3}{16} \div \dfrac{5}{8} =$

6. $25 \div \dfrac{5}{9} =$

7. $\dfrac{16}{35} \div \dfrac{2}{7} =$

8. $\dfrac{5}{24} \div \dfrac{3}{10} =$

9. $12 \div \dfrac{3}{8} =$

10. $\dfrac{14}{17} \div 21 =$

11. $\dfrac{3}{4} \div \dfrac{9}{32} =$

12. $\dfrac{12}{57} \div \dfrac{2}{3} =$

13. $\dfrac{5}{9} \div \dfrac{5}{18} =$

14. $\dfrac{2}{5} \div \dfrac{28}{31} =$

15. $\dfrac{5}{36} \div \dfrac{4}{9} =$

16. $\dfrac{9}{16} \div \dfrac{7}{18} =$

17. $42 \div \dfrac{18}{25} =$

18. $\dfrac{7}{8} \div \dfrac{5}{32} =$

Dividing Mixed Numbers

To divide mixed numbers, first write them as improper fractions. Then invert the divisor and multiply. Before multiplying, be sure to perform any cancellation possible. Rename the resulting quotient in its simplest terms.

Decimal fractions may be divided by common fractions and mixed numbers.

EXAMPLE

a. Find the quotient: $15\frac{3}{8} \div 5\frac{1}{4}$.

b. Find the quotient: $\$15.85 \div 3\frac{4}{9}$.

SOLUTION

a. $15\frac{3}{8} \div 5\frac{1}{4} = \frac{123}{8} \div \frac{21}{4} = \frac{\overset{41}{\cancel{123}}}{\underset{2}{\cancel{8}}} \times \frac{\overset{1}{\cancel{4}}}{\underset{7}{\cancel{21}}} = \frac{41}{14} = 2\frac{13}{14}$

b. $\$15.85 \div 3\frac{4}{9} = \frac{\$15.85}{1} \div \frac{31}{9}$

$\qquad = \frac{\$15.85}{1} \times \frac{9}{31} = \frac{\$142.65}{31} = \$4.601 \rightarrow \$4.60$

Self-Check
Find the quotient: $7\frac{2}{3} \div 4$.

Self-Check Answer
$1\frac{11}{12}$

Problems

Find the quotients in the following problems. Show the details of your work.

19. $32\frac{1}{6} \div 10\frac{2}{3} =$

20. $24\frac{3}{5} \div 8\frac{1}{5} =$

21. $5\frac{1}{8} \div 2\frac{2}{9} =$

22. $1\frac{2}{3} \div 1\frac{2}{3} =$

Answers

19._____

20._____

21._____

22._____

Business Applications 4.6

1. Environetics' design engineer has developed an energy-saving light requiring a platinum ribbon. The platinum ribbon comes from a roll 174 in. long. Each ribbon must be $\frac{3}{32}$ in. long. How many ribbons can be cut from the roll?

2. Rita Garland wishes to tile her 18-ft by 15-ft office with 9-in. by 9-in. tiles. How many whole tiles will she need to do the job?

3. Carlson Nursery has a garden plot measuring 162 ft by 162 ft. They plan to plant rows of seeds $\frac{3}{4}$ of a yard apart, planting as many rows as possible. How many rows will they be able to plant?

4. Betty Giddens works in the cutting rooms of an exclusive women's clothing manufacturer. Her next work order is a piece of fabric $128\frac{3}{8}$ yd long, from which she must cut skirts requiring $2\frac{1}{4}$ yd each. How many skirts will she be able to cut from this piece of fabric? (*Hint:* Round your answer down to the next lowest whole unit.)

5. The city of Alexandria is planning a $22\frac{1}{4}$-mile long bus route. If a bus stop is placed every $1\frac{1}{2}$ mi, how many stops will be on the route? (Round your answer to the nearest whole number.)

6. Canteen Cooperative serves soup as part of its daily menu. The soup is packed in cans containing 46 oz each; the average portion size is $7\frac{1}{3}$ oz. On an average day, 125 portions of soup are served. To the nearest whole can, how many cans of soup are used each day?

7. Richard Rodino paid $9,481.75 for the purchase of common stock. The purchase price was 48\frac{7}{8}$ a share. How many shares of stock did he purchase?

8. One package of natural cereal costs 63¢ for $7\frac{1}{2}$ oz. Another costs $1.39 for $18\frac{3}{4}$ oz.

a. What is the price of one ounce to the nearest tenth of a cent in the smaller package?

b. What is the price of one ounce to the nearest tenth of a cent in the larger package?

9. Lloyd King is a top sales representative for Global Safety Products. Last month his sales expenses amounted to $4,800, which represented $\frac{3}{32}$ of his sales for the month.

a. What was the total of Lloyd's sales for the month?

b. Lloyd's commission is $\frac{2}{15}$ of his net sales (total sales less expenses) for the month. What was the amount of his commission, to the nearest cent, for that same month?

Answers

6. _____

7. _____

8. a. _____

b. _____

9. a. _____

b. _____

Activities 4.6

Graph Activity

Mary Reese sells golf clubs for men, women, and youth. Her commission is $\frac{2}{17}$ of sales. If Mary sells $5,420 in one week, what is the amount of her commission? Round your answer to the nearest cent. _____

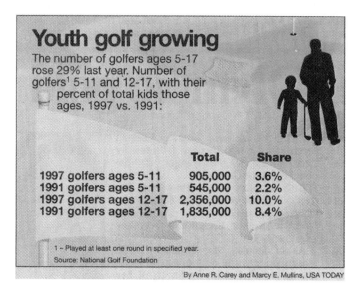

Youth golf growing

The number of golfers ages 5-17 rose 29% last year. Number of golfers[1] 5-11 and 12-17, with their percent of total kids those ages, 1997 vs. 1991:

	Total	Share
1997 golfers ages 5-11	905,000	3.6%
1991 golfers ages 5-11	545,000	2.2%
1997 golfers ages 12-17	2,356,000	10.0%
1991 golfers ages 12-17	1,835,000	8.4%

1 – Played at least one round in specified year.
Source: National Golf Foundation

By Anne R. Carey and Marcy E. Mullins, USA TODAY

Challenge Activity

Check your stocks from the problem in the Challenge Activity (page 200) one week after purchase date. Calculate the amount of gain/loss for each stock individually. Then add both stocks for a total gain/loss. What were the results?

Internet Activity

To divide fractions, just multiply. See how at **http://www.math.utep.edu/ sosmath/algebra/fraction/frac3/frac35/frac35.html**

SKILLBUILDER 4.7

Computing Fractional Parts

Learning Objectives

* Use fractional parts to solve problems.
* Use fractional parts to find the number of items when given unit price and total cost.

Fractional Parts

Any number that can be divided evenly into another number is known as a **fractional part** of that number. Many computations can be simplified by recognizing the fractional parts of one or more of the terms and by using those values in place of the original figures in the computation.

For example, suppose you want to determine how much it would cost to purchase 1,264 metric converter slides at 25¢ each. Rather than multiply 1,264 by 25¢, you should recognize that 25¢ equals $\frac{1}{4}$ of $1. If the items sold for $1 each, the cost would be $1,264. However, since they sold for 25¢, or $\frac{1}{4}$ of a dollar, you can multiply $1,264 by $\frac{1}{4}$ to obtain a purchase cost of $316.

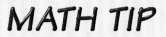

MATH TIP

Fractional parts make it possible to do many calculations mentally.

EXAMPLE

a. What part of $1, or 100¢ is $37\frac{1}{2}$¢?

b. Find the cost of 432 items at $37\frac{1}{2}$¢ each.

SOLUTION

a. Place $37\frac{1}{2}$¢ over $1.00 (or 100 cents) and rename the fraction in its simplest terms.

$$37\frac{1}{2} \div 100 = \frac{37\frac{1}{2}}{100} = \frac{2 \times 37\frac{1}{2}}{2 \times 100} = \frac{75}{200} = \frac{3}{8}$$

—or—

$$37\frac{1}{2} \div 100 = \frac{75}{2} \div \frac{100}{1} = \frac{\overset{3}{\cancel{75}}}{2} \times \frac{1}{\underset{4}{\cancel{100}}} = \frac{3}{8}$$

b. $37\frac{1}{2}$¢ $= \frac{3}{8}$ of $1

$$\frac{\overset{54}{\cancel{432}}}{1} \times \frac{3}{\underset{1}{\cancel{8}}} = \$162$$

Self-Check Answers

$\frac{1}{8}$

$471

Self-Check

a. What part of 44 is $5\frac{1}{2}$?

b. Find the cost of 1,256 items at $37\frac{1}{2}$¢ each.

Name:_____ Date:_____

Problems

Find the fractional parts of $1 represented by the following amounts. Show details of your work.

1. $50¢ = $ _____ of $1

2. $6\frac{1}{4}¢ = $ _____ of $1

3. $8\frac{1}{3}¢ = $ _____ of $1

4. $33\frac{1}{3}¢ = $ _____ of $1

5. $87\frac{1}{2}¢ = $ _____ of $1

6. $37\frac{1}{2}¢ = $ _____ of $1

7. $25¢ = $ _____ of $1

8. $12\frac{1}{2}¢ = $ _____ of $1

9. $75¢ = $ _____ of $1

10. $16\frac{2}{3}¢ = $ _____ of $1

11. $83\frac{1}{3}¢ = $ _____ of $1

12. $62\frac{1}{2}¢ = $ _____ of $1

Answers

1._____
2._____
3._____
4._____
5._____
6._____
7._____
8._____
9._____
10._____
11._____
12._____

Student Success Hints

Use the split-page method of note taking. Split the page with a third of the page as a left margin. Take notes on the right side of the page and use the left side for processing/learning—write companion questions (questions over your notes), terms, work out formulas, etc.

© Glencoe/McGraw-Hill

Skillbuilder 4.7 Computing Fractional Parts **209**

Answers

13. _____

14. _____

15. _____

16. _____

17. _____

18. _____

13. 408 qt @ $16\frac{2}{3}$¢

14. 368 in @ $6\frac{1}{4}$¢

15. 735 in @ $33\frac{1}{3}$¢

16. 252 in @ $12\frac{1}{2}$¢

17. 1,272 qt @ $66\frac{2}{3}$¢

18. 492 qt @ $8\frac{1}{3}$¢

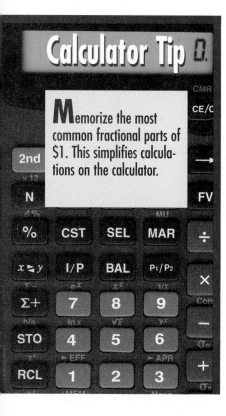

Calculator Tip

Memorize the most common fractional parts of $1. This simplifies calculations on the calculator.

Finding the Number of Items

Fractional parts can be used to find the quantity of items bought when the unit price and the total cost are known.

EXAMPLE

Jean McCrea, a candidate for public office, spent $108 for monogrammed pencils to be given to prospective voters. The pencils cost $12\frac{1}{2}$¢ each. How many pencils were bought?

SOLUTION

$12\frac{1}{2}$¢ $= \dfrac{1}{8}$ of $1 $108 $\div \dfrac{1}{8} = \dfrac{108}{1} \times \dfrac{8}{1} = 864$ pencils

 Self-Check Answer
232 yd

 Self-Check
Cotton cloth costs $87\frac{1}{2}$¢ a yard. If a total purchase came to $203, how many yards were bought?

Problems

19. $70 for decals at $8\frac{1}{3}$¢ each

20. $213.50 for paint thinner at $87\frac{1}{2}$¢ a pint

21. $135 for felt at $93\frac{3}{4}$¢ a meter

22. $168 for rice flour at 50¢ a pound

23. $225 for orange juice at 75¢ a quart

Business Applications 4.7

1. Serena Rolfe, owner of Rolfe's Auto Repair, purchased automatic transmission fluid for her shop storeroom. Ms. Rolfe paid $1.25 a quart for automatic transmission fluid. Her total bill came to $15.00. How many quarts did she purchase?

2. Tariq Keller is in charge of purchasing supplies for High Performance Auto Supply Store. When reordering brake fluid, Mr. Keller paid 87.5¢ a pint. The total bill came to $71.75. How many pints did he purchase?

3. During monthly inventory at Routh's Pharmacy, Ms. Routh decided she needed to restock her shelves with vitamin C tablets. She paid $90.00 for a case of tablets. If each bottle cost $3.75, how many bottles did she purchase?

Activities 4.7

Graph Activity

If a teen's family spends $225 at the grocery store, what part of that amount will be directed by the teen? Show your answer in dollars, as a fraction, and as a percent.

dollars   _____

fraction   _____

percent   _____

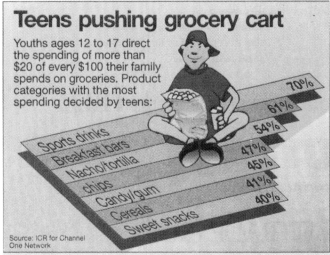

Teens pushing grocery cart

Youths ages 12 to 17 direct the spending of more than $20 of every $100 their family spends on groceries. Product categories with the most spending decided by teens:

Sports drinks — 70%
Breakfast bars — 61%
Nacho/tortilla chips — 54%
Candy/gum — 47%
Cereals — 45%
Sweet snacks — 41%
— 40%

Source: ICR for Channel One Network

By Anne R. Carey and Web Bryant, USA TODAY

Challenge Activity

Mr. Brian has contacted you, as an assistant for Blankenship Financial Services, to choose and purchase $25,000.00 worth of stocks from NASDAQ/National Market System Composite. Your challenge is to choose four (4) stocks (other than the stocks chosen in the previous exercises) from the newspaper and to locate a broker in your community to determine the broker's commission and any other investment fees. Remember to round to the nearest cent. Name the financial service, list the fees, and the amount charged for any fees. What is the closing price per share for each stock? What is the total cost for each stock separately? What is the total amount spent?

Internet Activity

What shape are your fractions in? Take a quiz of your visual ability to compute fractional parts at

http://math.rice.edu/~lanius/Patterns/

SKILLBUILDER 4.8

Using Fractional Parts

Learning Objective

- **Use fractional parts to speed up computations.**

Using Fractional Parts

Fractional parts are useful only if their values and relationships are recognized quickly. The following list shows the fractional parts of $1 that are used most frequently. Use it as necessary for the problems.

$\$\dfrac{1}{2} = 50\cent$ $\$\dfrac{1}{5} = 20\cent$ $\$\dfrac{1}{8} = 12\tfrac{1}{2}\cent$ $\$\dfrac{7}{10} = 70\cent$

$\$\dfrac{1}{3} = 33\tfrac{1}{3}\cent$ $\$\dfrac{2}{5} = 40\cent$ $\$\dfrac{3}{8} = 37\tfrac{1}{2}\cent$ $\$\dfrac{9}{10} = 90\cent$

$\$\dfrac{2}{3} = 66\tfrac{2}{3}\cent$ $\$\dfrac{3}{5} = 60\cent$ $\$\dfrac{5}{8} = 62\tfrac{1}{2}\cent$ $\$\dfrac{1}{12} = 8\tfrac{1}{3}\cent$

$\$\dfrac{1}{4} = 25\cent$ $\$\dfrac{4}{5} = 80\cent$ $\$\dfrac{7}{8} = 87\tfrac{1}{2}\cent$ $\$\dfrac{1}{16} = 6\tfrac{1}{4}\cent$

$\$\dfrac{3}{4} = 75\cent$ $\$\dfrac{1}{6} = 16\tfrac{2}{3}\cent$ $\$\dfrac{1}{10} = 10\cent$ $\$\dfrac{1}{20} = 5\cent$

$\$\dfrac{5}{6} = 83\tfrac{1}{3}\cent$ $\$\dfrac{3}{10} = 30\cent$

Fractional parts can be used to speed up computations for values greater than $1 too.

> ### Student Success Hints
>
> Listen and record main ideas in your notes. Take note of topics the instructor emphasizes for problem solving or for the test. Use phrases instead of complete sentences, use abbreviations and symbols to help get the information down faster, and leave space between ideas.

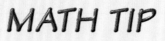

MATH TIP

Memorizing the fractional parts you use most frequently will save you time when calculating fractions.

 **EXAMPLE**

Find the total cost.

a. 320 pieces at $1.12\frac{1}{2}$ each.

b. 256 units at $37.50 each.

SOLUTION

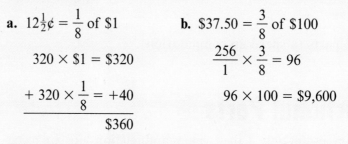

a. $12\frac{1}{2}¢ = \dfrac{1}{8}$ of $1

$$320 \times \$1 = \$320$$

$$+\ 320 \times \dfrac{1}{8} = +40$$

$$\overline{\hspace{2cm}\$360}$$

b. $37.50 = \dfrac{3}{8}$ of $100

$$\dfrac{256}{1} \times \dfrac{3}{8} = 96$$

$$96 \times 100 = \$9{,}600$$

 Self-Check Answer $700

 Self-Check

Find the cost of 112 items at $6.25 each.

Name:_____   Date:_____

Problems

Find the total cost of each of the following groups of purchases. In the parentheses after each price, show what fractional part of $1 that price represents. Round your answers to the nearest cent where necessary.

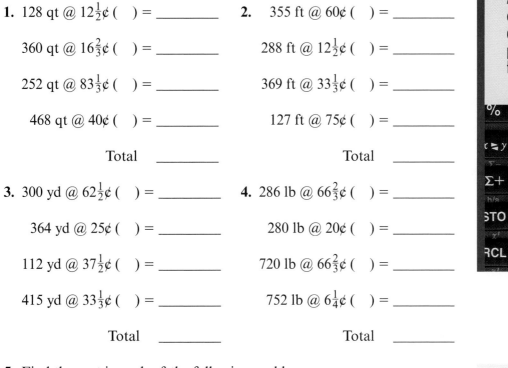

1. 128 qt @ $12\frac{1}{2}$¢ (   ) = _____

    360 qt @ $16\frac{2}{3}$¢ (   ) = _____

    252 qt @ $83\frac{1}{3}$¢ (   ) = _____

    468 qt @ 40¢ (   ) = _____

            Total _____

2. 355 ft @ 60¢ (   ) = _____

    288 ft @ $12\frac{1}{2}$¢ (   ) = _____

    369 ft @ $33\frac{1}{3}$¢ (   ) = _____

    127 ft @ 75¢ (   ) = _____

            Total _____

3. 300 yd @ $62\frac{1}{2}$¢ (   ) = _____

    364 yd @ 25¢ (   ) = _____

    112 yd @ $37\frac{1}{2}$¢ (   ) = _____

    415 yd @ $33\frac{1}{3}$¢ (   ) = _____

            Total _____

4. 286 lb @ $66\frac{2}{3}$¢ (   ) = _____

    280 lb @ 20¢ (   ) = _____

    720 lb @ $66\frac{2}{3}$¢ (   ) = _____

    752 lb @ $6\frac{1}{4}$¢ (   ) = _____

            Total _____

5. Find the cost in each of the following problems.

    a. 320 items @ $37\frac{1}{2}$¢ (   )

    b. 560 items @ $62\frac{1}{2}$¢ (   )

    c. 200 items @ 5¢ (   )

    d. 550 items @ 80¢ (   )

Answers

5. a. _____

   b. _____

   c. _____

   d. _____

Business Applications 4.8

1. Compute the extensions on the following purchases using fractional parts for all values. In the parentheses after the unit price, show the fraction of $10, $100, and so on, that the price represents. Round to the nearest cent. Then find the total.

42 jackets at $62.50 each (   ) = _____

74 cotton shirts at $12.50 each (   ) = _____

116 pairs socks at $6.25 each (   ) = _____

15 pairs jeans at $25.00 each (   ) = _____

48 T-shirts at $8.33 each (   ) = _____

13 sport jackets at $90.00 each (   ) = _____

Total _____

2. The accountant for Tapes, Etc., Arleen's Art Shop, and Mechanic's World completes charts like the following to show sales for the month of May. Compute the extensions using fractional parts to determine monthly sales for each item. Show the fraction of $10, $100, and so on, that the price represents. Then find the totals.

TAPES, ETC.

18 radios at $37.50 each (   ) = _____

15 compact disc cases at $8.33 each (   ) = _____

12 tapes at $6.25 each (   ) = _____

23 compact discs at $8.75 each (   ) = _____

60 carrying cases at $8.33 each (   ) = _____

87 video tapes at $3.75 each (   ) = _____

Total _____

ARLEEN'S ART SHOP

46 brushes at $6.25 each (   ) = _____

48 paint boxes at $9.00 each (   ) = _____

32 cans paint thinner at $6.25 each (   ) = _____

24 clay pots at $16.67 each (   ) = _____

18 table easels at $8.75 each (   ) = _____

20 palettes at $12.50 each (   ) = _____

10 sheets sandpaper at $0.50 each (   ) = _____

Total   _____

MECHANIC'S WORLD

26 6-piece wrench sets at $37.50 each (   ) = _____

52 drills at $12.50 each (   ) = _____

318 screw packs at $6.25 each (   ) = _____

6 small socket sets at $37.50 each (   ) = _____

12 entrance locks at $16.67 each (   ) = _____

15 latch sets at $8.33 each (   ) = _____

10 channel locks at $6.25 each (   ) = _____

Total   _____

Activities 4.8

Graph Activity

Mary Clarke is a realtor. She gets $\frac{1}{2}$ of any commission that each of her associates make and $\frac{1}{4}$ of any commission that each of her senior associates make. At the end of the week, how much of these commissions did Mary receive? _____

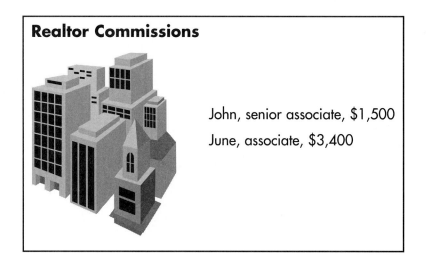

Realtor Commissions

John, senior associate, $1,500

June, associate, $3,400

Challenge Activity

Follow and calculate any changes in the stocks from Skillbuilder 4.7 (page 213), once a week for one month. Can you see a pattern of rise or fall? How much did each stock gain/lose? How much did your total investment gain/lose weekly and at the end of the month?

Internet Activity

The use of fractional parts is essential in the design of quilts. Here is a patch quilt pattern, see the pinwheel:

http://ares.redsword.com/dduperault/art/basic2.gif

What fractional parts are pieced together to make an advanced pinwheel design?

UNIT 5 Percents

Percents are probably one of the most common ways to show data because they give us a means of determining what part of the whole is involved. They also allow us to compare data from different-sized groups. This chart from *The Wall Street Journal* shows the results of a survey of how hourly and salaried workers view their jobs.

In this unit we study percents and their application to commissions and to markup and markdown. In addition, we study graphs, tables, ratios, and proportions.

HOW SALARIED AND HOURLY WORKERS VIEW THEIR JOBS

View Jobs	% Salaried	% Hourly
Saw a connection between good performance and pay increase	62	48
Were not satisfied with pay	47	55
Were not satisfied with benefits	36	46
Felt no job security	54	65
Just a place to work	54	67
Not a good place to work	42	53
Rated work environments negatively	35	46

MATH CONNECTIONS

Crew Manager

Susan Limes is a crew manager for Capital Builders. Susan is responsible for ordering supplies, scheduling work crews, and inspecting the workmanship to ensure that it meets federal, state, and local standards. When a problem arises, Susan must decide on a solution. Susan spends most of her work days traveling from one job site to another, making sure that all of the construction projects are progressing according to schedule.

Math Application

Susan submits daily time sheets to her central office. The time sheet lists each job site, the time Susan spent at the site, and Susan's total hours for the day. Following is Susan's time sheet for Monday, February 7.

Job Site	Hours at Site
Long St.	$1\frac{1}{2}$
Main St.	$2\frac{1}{4}$
Franklin Ave.	$3\frac{1}{2}$

Total hours on site $7\frac{1}{4}$

Total hours, including travel time ($2\frac{3}{4}$ hrs)   10.0

What percent of the day did Susan spend at each job? Divide the time spent at each job site by the total hours on site. Round to the nearest percent.

Long St. $1\frac{1}{2} \div 7\frac{1}{4} =$ _____ = _____

Main St. $2\frac{1}{4} \div 7\frac{1}{4} =$ _____ = _____

Franklin Ave. $3\frac{1}{2} \div 7\frac{1}{4} =$ _____ = _____

Critical Thinking Problem

Susan Limes is scheduling work crews for the next two weeks. How might she project her total costs for the owners?

SKILLBUILDER 5.1

Using Percents and the Percentage Formula

Learning Objectives

* **Rename percents as decimals and fractions.**
* **Rename decimals and fractions or mixed numbers as percents.**
* **Use the percentage formula to find the percentage.**

Renaming Percents as Decimals and Fractions

Percent means *per hundred.* Percents can be expressed as decimals or as common fractions or mixed numbers. To rename a percent as a decimal, remove the percent sign and divide the percent by 100 (or move the decimal point two places to the left).

To rename a percent as a common fraction, first express the percent as a decimal. Then rename the decimal as a common fraction reduced to lowest terms.

EXAMPLE

a. Rename 57% as a decimal.
b. Rename $8\frac{3}{4}\%$ and $128\frac{2}{5}\%$ as common fractions reduced to lowest terms

SOLUTION

a. $57\% = 0\underset{\frown}{57}. = 0.57$

b. $8\frac{3}{4}\% = 0.0875 = \dfrac{875}{10,000} = \dfrac{7}{80}$

c. $128\frac{2}{5}\% = 128.4 = 1.284 = 1\dfrac{284}{1,000} = 1\dfrac{71}{250}$

Self-Check
a. Rename 142% as a decimal.
b. Rename 150% as a common fraction or mixed number in lowest terms.

MATH TIP

The number of decimal places indicates the power of 10 in the denominator when you rename a percent as a fraction. Always reduce the fraction to lowest terms.

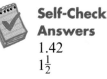
Self-Check Answers
1.42
$1\frac{1}{2}$

Answers

1. _____
2. _____
3. _____
4. _____
5. _____
6. _____
7. _____
8. _____
9. _____

Problems

Rename each percent as a decimal.

1. 24% **2.** 93% **3.** 17%

4. 8% **5.** 20.2% **6.** 158%

7. 0.5% **8.** 200% **9.** 1.15%

Rename each percent as a fraction or mixed number. Reduce to lowest terms, where necessary.

10. 25% = _____ **11.** 35% = _____ **12.** 16% = _____

13. 7% = _____ **14.** 45% = _____ **15.** $33\frac{1}{3}$% = _____

16. 15% = _____ **17.** $62\frac{1}{2}$% = _____ **18.** $6\frac{1}{4}$% = _____

19. 225% = _____ **20.** $83\frac{1}{3}$% = _____ **21.** $37\frac{1}{2}$% = _____

22. $2\frac{1}{4}$% = _____ **23.** $\frac{1}{4}$% = _____ **24.** 375% = _____

25. $87\frac{1}{2}$% = _____

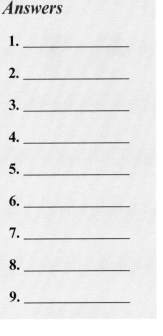

Renaming Decimals and Fractions as Percents

To rename a decimal as a percent, multiply the decimal by 100 and add a percent sign.

To rename a fraction as a percent, rename the fraction as a decimal by dividing the numerator by the denominator, and then rename the decimal as a percent.

EXAMPLE

a. Rename 0.333 as a percent.

b. Rename $\frac{1}{4}$ and $1\frac{1}{8}$ as percents.

SOLUTION

a. $0.333 = 0.333\% = 33.3\%$

b. $\frac{1}{4} = 0.25 = 0.25\% = 25\%$

$1\frac{1}{8} = 1.125 = 1.125\% = 112.5\%$

Self-Check Answers
45%
$62\frac{1}{2}$%

Self-Check

a. Rename 0.45 as a percent.

b. Rename $\frac{5}{8}$ as a percent

Problems

Rename each common fraction as (a) a decimal fraction (rounded off to the nearest ten thousandth where necessary) and then to (b) a percent.

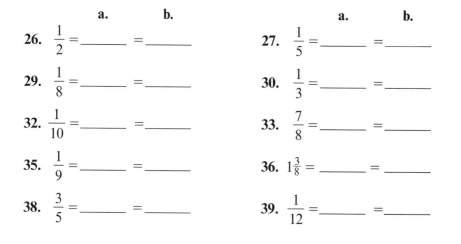

	a.	b.		a.	b.		a.	b.
26. $\frac{1}{2}$ =	____	____	27. $\frac{1}{5}$ =	____	____	28. $\frac{3}{8}$ =	____	____
29. $\frac{1}{8}$ =	____	____	30. $\frac{1}{3}$ =	____	____	31. $\frac{4}{5}$ =	____	____
32. $\frac{1}{10}$ =	____	____	33. $\frac{7}{8}$ =	____	____	34. $\frac{1}{6}$ =	____	____
35. $\frac{1}{9}$ =	____	____	36. $1\frac{3}{8}$ =	____	____	37. $\frac{2}{3}$ =	____	____
38. $\frac{3}{5}$ =	____	____	39. $\frac{1}{12}$ =	____	____	40. $\frac{1}{16}$ =	____	____

Using the Percentage Formula

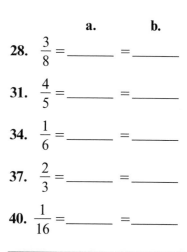

The **percentage formula** is

$$\text{Base} \times \text{Rate} = \text{Percentage}$$

A percent is also called a **rate.** When a rate is multiplied by a number, called the **base,** the result is a **percentage.** Use the fractional equivalent for the rate (percent) whenever it will help solve the problem.

Calculator Tip

If you misplace the decimal point, your percentage will be incorrect. Estimate your answer to check if it is logical.

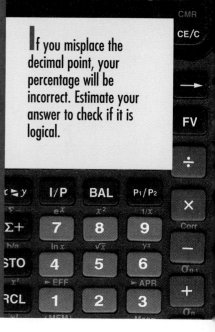

EXAMPLE

What is $62\frac{1}{2}\%$ of 120?

SOLUTION

$62\frac{1}{2}\% = 0.625$

Base × Rate = Percentage
120 × 0.625 = 75

Self-Check
What is 30% of $615?

Self-Check Answer
$184.50

41. _____

42. _____

43. _____

44. _____

45. _____

46. _____

47. _____

48. _____

49. _____

50. _____

51. _____

52. _____

53. _____

54. _____

55. _____

56. _____

57. _____

58. _____

59. _____

60. _____

61. _____

Problems

Compute the percentage in the following problems using the formula
base × rate = percentage ($B \times R = P$). Use the fractional equivalent for the rate (percent) whenever it will help solve the problem.

41. 8% of $500

42. 5% of $42.80

43. 25% of $5,800

44. 2% of 90

45. 72% of 613

46. $8\frac{1}{3}$% of 288

47. $12\frac{1}{2}$% of $8,010

48. $2\frac{1}{2}$% of $340

49. $16\frac{2}{3}$% of 264

50. $37\frac{1}{2}$% of 4,260

51. 9% of 1,854

52. $31\frac{1}{4}$% of $2,500

53. 52% of 18,500

54. $11\frac{1}{9}$% of 810

55. $7\frac{1}{2}$% of $7,800

56. $62\frac{1}{2}$% of 9,000

57. $41\frac{3}{4}$% of 864

58. $87\frac{1}{2}$% of 12,800

59. $43\frac{3}{4}$% of 795

60. $33\frac{1}{3}$% of $7,218

61. 21% of 6,320

Name:_____ Date:_____

Business Applications 5.1

Answers

1. _____

2. a. _____

   b. _____

   c. _____

3. a. _____

   b. _____

1. Last month, Roberta Jameson was paid a commission of $7\frac{1}{2}\%$ on sales totaling $78,000. How much did she earn for the month?

2. Albert Graham received an $8\frac{1}{3}\%$ increase in his $31,500 annual salary.

   a. What was the total amount of his increase?

   b. How much did his monthly salary increase?

   c. What is his new annual salary?

3. Emma Channing, sales manager for Queens' Cosmetics, is projecting an $11\frac{1}{9}\%$ increase in sales for the next quarter. Sales for this quarter were $1,459,008.

   a. What is the expected amount of increase for the next quarter's sale?

   b. If the increase is attained, what will total sales be for the next quarter?

4. José Martinez, marketing manager for Queens' Cosmetics, conducted a market survey and determined that 37.5% of the 2,150 people questioned recognized at least one product produced by Queens' Cosmetics. How many people surveyed knew about their products?

5. Fifteen percent of the new subscribers of Sports News Magazine pay for their subscription by check. If there are 4,000 new subscribers, how many paid by check?

6. An urgent care center in South Carolina reported that stress-related illnesses were reported in 30% of the patients they treated during the month of December. If they treated 870 patients during December, how many had stress-related illnesses?

7. Aiken Chiropractic increased last year's revenue by $37\frac{1}{2}\%$. If last year's revenue was $320,800, by what amount did revenue increase?

Activities 5.1

Graph Activity

Look at the graph below. If there are 98,000,000 men traveling this summer, how many men will travel 100 miles to visit a beach? _____

Who's on the beach

Nearly 50 million adults — 35% of travelers — will travel 100 miles or more one way to visit a beach this summer. Percent of each of these groups planning a beach trip:

Men	41%
Women	30%
Age 18-34	45%
Age 35-54	38%
Age 55-up	20%
Child at home	39%
No children	32%

Source: Travel Industry Association   By Anne R. Carey and Elys A. McLean, USA TODAY

Challenge Activity

Josie McGill established a personal services business two years ago, providing services to homebound people to do their shopping, accompany them shopping, to the doctor, or on other personal errands. Business has prospered, and she now employs three assistants. Josie took out a business loan to upgrade her business equipment and facilities. The upgrade encompasses improvement of computer facilities for \$3,000, which is $\frac{1}{5}$ of the loan. Down payment on the lease of an additional van will use 25 percent of the loan. A new telephone system will use $\frac{1}{7}$ of the loan, and pagers for everyone will use 1 percent of the loan. What was the amount of the loan, and how much did she spend in all on the upgrades?

Internet Activity

The economy has upswings and downturns. These trends are often expressed as percents. This article discusses business success in the counties of Washington State. Rename several of these percents as fractions.

wysiwyg://http://www.wa.gov/misc/q197qbr.html

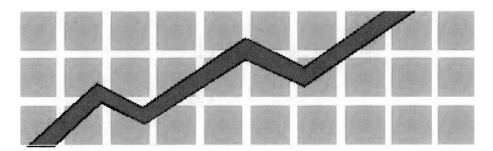

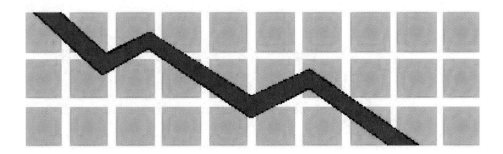

SKILLBUILDER 5.2

Computing the Rate, Base, Increase, and Decrease

Learning Objectives

- **Use the percentage formula to compute base and rate.**
- **Compute either the amount or the rate of increase.**
- **Compute either the amount or the rate of decrease.**

Using The Percentage Formula

The basic percentage formula can be used to find either the base or the rate. The following diagram is useful.

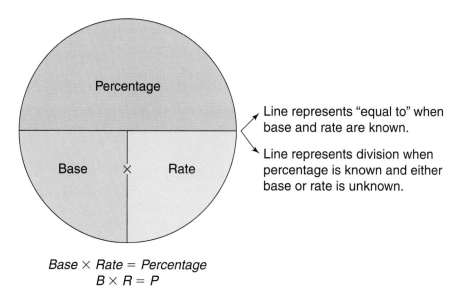

Line represents "equal to" when base and rate are known.

Line represents division when percentage is known and either base or rate is unknown.

Base × *Rate* = *Percentage*
$B \times R = P$

Cover the word representing the unknown element in the formula. The remaining portions of the chart indicate what operation is necessary to find the unknown. For example, to find 30% of 340, cover the word *percentage* since it is unknown. The diagram shows that you should multiply the base times the rate.

If the base is unknown, cover the word *base*. The diagram shows that you divide the percentage by the rate. If the rate is unknown, divide the percentage by the base.

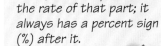

Self-Check Answers
250
$12\frac{1}{2}\%$

EXAMPLE

a. 102 is 30% of what number?
b. What percent of 520 is 130?

SOLUTION

a. $P \div R = B$
  $102 \div 0.30 = 340$
b. $130 \div 520 = 0.25 = 25\%$

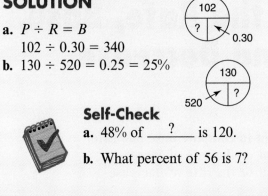

Self-Check

a. 48% of _____?_____ is 120.

b. What percent of 56 is 7?

Name:_____ Date:_____

Problems

Find the base in each of the following problems. Remember to rename the percent as a decimal or fraction.

1. 25% of $__?__ = $144

2. 30% of $__?__ = $21.90

3. 40% of $__?__ = $75.60

4. $12\frac{1}{2}$% of $__?__ = $42

5. 8% of $__?__ = $4.96

6. $16\frac{2}{3}$% of $__?__ = $264

7. $37\frac{1}{2}$% of $__?__ = $18

8. $\frac{1}{2}$% of $__?__ = $10

9. $62\frac{1}{2}$% of $__?__ = $55

10. $8\frac{1}{3}$% of $__?__ = $9

Find the rate, to the nearest hundredth, where necessary, in each of the following problems.

11. $18 is what percent of $72?

12. What percent of 128 is 3.84?

13. $78 is what percent of $156?

14. $13.50 is what percent of $162?

Answers

1._____
2._____
3._____
4._____
5._____
6._____
7._____
8._____
9._____
10._____
11._____
12._____
13._____
14._____

Answers

15. _____

16. _____

17. _____

18. _____

19. _____

20. _____

15. 64 is what percent of 192?

16. ___?___ % of $60 is $6

17. ___?___ % of 150 = 30

18. ___?___ % of 63 = $7\frac{7}{8}$

19. ___?___ % of $64 = $56

20. ___?___ % of $114 = $38

Increase

An **increase problem** is one in which the new amount (percentage) is larger than the original amount (base). Adding 100% (representing the base) to the rate of increase results in a new rate that represents the new amount (percentage). When the base is multiplied by this new rate, the resulting product will be the new percentage.

When the percentage and the rate of increase are known, the base is found by dividing the percentage by 100% plus the rate of increase.

Student Success Hints

Fill in any missing information the same day that you take notes. Spell out any important abbreviations. If you cannot recall missing information, check your textbook or another student's notes.

EXAMPLE

a. The price of coffee has just increased by 20% from $2.15 a pound. What is the increased price?

b. Coffee is now selling at $2.58 per pound, which is 20% greater than its earlier price. What was the price of coffee before the increase?

SOLUTION

a. B × R = P
100% represents the current amount.

$$\begin{array}{c} Original \\ Amount \end{array} \times \left(100\% + \begin{array}{c} Rate\ of \\ Increase \end{array}\right) = \begin{array}{c} New\ Larger \\ Amount \end{array}$$

$2.15 × (100% + 20%) = ?
$2.15 × 1.20 = $2.58

b. The base is unknown. Divide the percentage ($2.58) by the rate (100% + 20% increase) to find the original price.

$$\begin{array}{ccccc} P & ÷ & R & = & B \end{array}$$
$$\begin{array}{c} New\ Larger \\ Amount \end{array} ÷ \left(100\% + \begin{array}{c} Rate\ of \\ Increase \end{array}\right) = \begin{array}{c} Original \\ Amount \end{array}$$

$2.58 ÷ (100% + 20%) = ?
$2.58 ÷ 1.20 = $2.15

Check
$2.15 × 0.20 = $0.43 $2.15 + $0.43 = $2.58

Problems

Find the percentage in each problem.

21. 120 increased by 10%

22. 56 increased by 25%

23. $6.80 increased by 20%

24. 50 increased by 100%

Find the base of each problem.

25. ___?___ × 1.25 = 60

26. ___?___ × 1.6 = 80

27. ___?___ × 1.20 = 240

28. ___?___ × 2.00 = 356

29. ___?___ increased by 20% = 120

30. ___?___ increased by 60% = 32

31. ___?___ increased by $12\frac{1}{2}$ = 99

32. ___?___ increased by 100% = 78

Answers

21._____
22._____
23._____
24._____
25._____
26._____
27._____
28._____
29._____
30._____
31._____
32._____

Calculator Tip

When finding rate or base, use the percent key instead of the equals key to multiply or divide by a percent.

Find the rate in each problem.

33. 424 increased by _____ = 477

34. 115 increased by _____ = 138

35. 79 increased by _____ = 158

Decrease

With **decrease problems,** the percentage—the new amount—is always less than the base. Subtracting the rate of decrease from 100% (representing the base) results in a new rate. This new rate is also known as the complement of the original rate. When the base is multiplied by this new rate, the resulting product is the new percentage.

When the percentage and the rate of decrease are known, the base is found by dividing the percentage by 100% minus the rate of decrease, or the complement of the rate of decrease.

When the percentage and rate of decrease are known, to find the base (the original amount)

- Subtract the rate of decrease from 100%
- This is the complement of the decrease
- Divide the percentage by the complement (100% − rate of decrease)

EXAMPLE

A portable CD player is priced at $208.50, which is a 25% reduction from the original price. What was the original price?

SOLUTION

P	$\div$	R		$=$	B
New Lesser Amount	$\div$ (100%	−	Rate of Decrease)	$=$	Original Amount
$208.50	$\div$ Complement (100%	−	25%)	$=$	?
$208.50	$\div$ 0.75			$=$	$278

P	$\div$	R		$=$	B
New Lesser Amount	$\div$ (100%	−	Rate of Decrease)	$=$	Original Amount
$208.50	$\div$ Complement (100%	−	25%)	$=$	?
$208.50	$\div$ 0.75			$=$	$278

Check
$278 − $69.50 = $208.50

Name:_____ Date:_____

Problems

Find the percentage in each of the following decrease problems. Round percents to four decimal places.

36. 40 decreased by 25%

37. $120 reduced by $33\frac{1}{3}$%

38. $480 reduced by 16%

39. 978 reduced by 50%

40. 58 reduced by $37\frac{1}{2}$%

41. $129 reduced by $16\frac{2}{3}$%

Find the base in each decrease problem. Round percents to four decimal places.

42. ___?___ × 0.78 = 78

43. ___?___ × 0.625 = 640

44. ___?___ × 0.45 = 33.75

45. ___?___ × 0.50 = 1,025

46. ___?___ decreased by 29% = $177.50

47. ___?___ decreased by 15% = 85

Answers

36. _____

37. _____

38. _____

39. _____

40. _____

41. _____

42. _____

43. _____

44. _____

45. _____

46. _____

47. _____

48. ____?____ decreased by $8\frac{1}{3}\%$ = \$44

49. ____?____ decreased by 60% = 48

50. ____?____ decreased by $33\frac{1}{3}\%$ = 950

51. ____?____ decreased by 62.5% = 642

Find the rate in each problem.

52. 128 decreased by _____ = 102.4

53. 90 decreased by _____ = 44.1

54. 320 decreased by _____ = 200

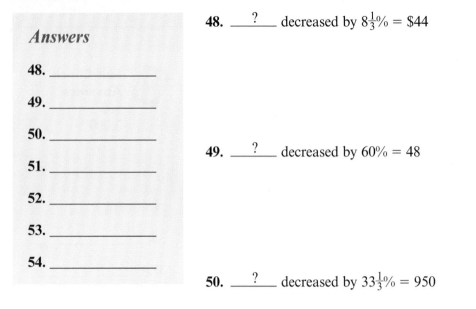

Calculator Tip

If your calculator does not have a percent key, change the percent to its decimal equivalent to find a percentage.

Business Applications 5.2

In the following problems, round percents to two decimal places where necessary.

Answers

1. _____

2. _____

3. _____

4. _____

1. Barthow's Pharmacy sells Plus Screen Lotion for $4 after a 20% discount. What was the original price?

2. Orbit Sales mistakenly included 5% sales tax in a bill to a tax-exempt organization. The invoice showed a total of $247.80, which included the 5% sales tax. What is the amount of tax that must be refunded?

3. Membership in the Four Seasons Racquet Club has dropped by $12\frac{1}{2}$%, according to Natashia Ling, the manager. This year's membership is 2,198 adults and children. What was last year's membership?

4. Juan contributes 5% of his gross pay to a retirement fund. Last month he contributed $150. What was Juan's gross pay for last month?

Activities 5.2

Graph Activity

If 1,500 men and women were surveyed, how many of each feel guilty? _____

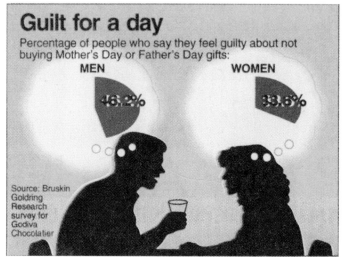

Guilt for a day

Percentage of people who say they feel guilty about not buying Mother's Day or Father's Day gifts:

MEN   46.2%

WOMEN   33.6%

Source: Bruskin Goldring Research survey for Godiva Chocolatier

By Cindy Hall and Gary Visgaitis, USA TODAY

Challenge Activity

Global Enterprises has major facilities in the United States, Europe, the Far East, and Latin America. Forty-two percent of their facilities are located in the United States, 26 percent in Europe, 15 percent in Asia, and 17 percent in Latin America. Because of worldwide economic conditions, Global has decided that it must reduce its total work force by 4,800 employees, which is 3 percent of its total work force. Of the 4,800 employees, 10 percent of the reduction will be made in U.S. facilities, 26 percent in European facilities, 43 percent in Asian facilities, and 21 percent in Latin American facilities. After the reduction, how many employees will remain in each of the four operational areas?

Internet Activity

You can compute the amount of dollars you save if you know the original price and the percent of reduction. Practice computing money saved at the following web site.

http://www.cne.gmu.edu/modules/dau/algebra/fractions/frac6_bdy.ht

SKILLBUILDER 5.3

Using Tables

Learning Objective

- **Use tables to find information needed to solve problems.**

Using Tables

A **table** is a systematic arrangement of information, usually numeric. The data are arranged in labeled rows and columns. Tables are read by selecting the row and column containing the information sought. The data required are found at the intersection of the row and the column. Tables are often used in business to provide a clear and concise presentation of data.

The table on the next page presents population data for the top 10 of the 150 largest metropolitan statistical areas (MSA). Past data show the percent of population change from 1980 to 1990; actual population in 1990; 1990–1997 estimated percent of change in population; 1997 estimated population; 1997–2001 projected percent of change in population; and 2001 projected population. The MSAs are ranked by population according to their 2001 projected figures. A Metropolitan Statistical Area, or MSA, is an area defined by the Office of Management and Budget (OMB) for use in publishing census data and other government statistics.

The information contained in a table may be precise, requiring a detailed listing of data, or it may be an approximation, in which the data are shown in a rounded format, as discussed in Skillbuilder 1.8.

Whether a precise format or a rounded format is to be used may be determined by the accuracy needed, and/or the amount of space available for the table or graph display. For example, in the table on page 244, the 1990 census of population in Los Angeles was 8,863,164. If precise accuracy were not required, or space did not permit the use of the full number, the population count could be abbreviated to reflect millions rounded to, say, three decimal places, as 8.863 million, or 8.9 million rounded to tenths.

POPULATION TRENDS, 1980 TO 2001

	1980 Rank	Past 1980–1990 % Change	1990 Census	1991 Rank	Present 1990–1997 % Change	1997 Estimate	Future 1997–2001 % Change	2001 Projection	2001 Rank
Los Angeles-Long Beach, CA	2	18.5	8,863,164	1	6.2	9,410,300	1.5	9,550,700	1
New York, NY	1	3.3	8,546,846	2	1.3	8,654,200	0.3	8,677,300	2
Chicago, IL	3	2.3	7,410,858	3	5.4	7,813,000	2.7	8,020,800	3
Boston-Worchester-Lawrence-Lowell-Brockton, MA-NH	6	6.6	5,685,998	7	2.4	5,820,700	3.0	5,994,400	4
Philadelphia, PA-NJ	4	2.9	4,922,175	4	0.8	4,959,400	0.3	4,973,800	5
Washington, DC-MD-VA-WV	7	21.4	4,223,485	6	8.3	4,575,400	5.0	4,805,600	6
Detroit, MI	5	−2.8	4,266,654	5	1.4	4,324,900	1.3	4,381,000	7
Houston, TX	8	20.6	3,322,025	8	14.6	3,807,400	7.1	4,079,300	8
Atlanta, GA	13	32.5	2,959,950	9	21.0	3,582,200	12.7	4,036,700	9
Dallas, TX	15	30.2	2,676,248	11	15.0	3,077,900	8.3	3,334,200	10

Source: Rand McNally Commercial Atlas, 1998

MATH TIP

Memorizing commonly used fraction-percent equivalents can be helpful when using tables.

EXAMPLE

Estimate the population of Atlanta in 1980.

SOLUTION

From 1980 to 1990, the population of Atlanta grew by 32.5%. Thus, the 1990 population was about 133% of the 1980 population. We can round the 1990 population to 3,000,000.

$133\% \times 1980 \text{ population} = 3,000,000$

$1980 \text{ population} = \dfrac{3,000,000}{1.33} = 2,255,639 = 2,250,000$

Self-Check Answer
3

Self-Check
About how many times as great as the population of Dallas is the population of Los Angeles?

Problems

1. List the cities whose population ranking remained the same for 1980, 1991, and 2001.

2. a. Which cities dropped in rank by one or more over the three periods shown?

   b. Which cities rose in rank by one or more over the three periods shown?

3. a. Which city has the greatest change in rank from 1980 to 2001?

   b. Which city has the greatest amount of population change projected from 1990 to 2001?

   c. What is the projected amount of population change for this city from 1990 to 2001?

   d. What is the percent of population increase or decrease for this city from 1990 to 2001? (*Note:* The percent of increase or decrease is found by comparing the change in population to the original population.)

Answers

1. _____

   _____

2. a. _____

   _____

   _____

   b. _____

   _____

   _____

   _____

   _____

3. a. _____

   b. _____

   c. _____

   d. _____

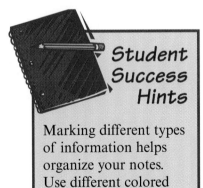

Student Success Hints

Marking different types of information helps organize your notes. Use different colored highlight pens to mark key terms, problem-solving tips, equations, and other information.

FOOD—RETAIL PRICES OF SELECTED ITEMS: 1990 TO 1996 (IN DOLLARS PER POUND, EXCEPT AS INDICATED. AS OF DECEMBER)

Food	[1990]	[1991]	[1992]	[1993]	[1994]	[1995]	[1996]
Rice, White, lg. grain, raw	0.49	0.51	0.53	0.50	0.53	0.55	0.55
Spaghetti and macaroni	0.85	0.86	0.86	0.84	0.87	0.88	0.84
Ground Chuck, 100% beef	2.02	1.93	1.91	1.91	1.84	1.85	1.85
Chicken, fresh, whole	0.86	0.86	0.88	0.91	0.90	0.94	1.00
Chicken breast, bone in	2.00	2.02	2.08	2.17	1.91	1.95	2.09
Turkey, frozen, whole	0.96	0.91	0.93	0.95	0.98	0.99	1.02
Eggs, Grade A, large (dozen)	1.00	1.01	0.93	0.87	0.87	1.16	1.31
Milk, fresh, whole (1/2 gal.)	1.39	1.40	1.39	1.43	1.44	1.48	1.65
Ice Cream (1/2 gal.)	2.54	2.63	2.49	2.59	2.62	2.68	2.94
Apples, red delicious	0.77	0.86	0.76	0.78	0.72	0.83	0.89
Bananas	0.43	0.42	0.40	0.41	0.46	0.45	0.48
Pears, anjou	0.79	0.88	0.80	0.89	(NA)	(NA)	1.06
Potatoes, white	0.32	0.28	0.31	0.36	0.34	0.38	0.33
Lettuce, iceberg	0.58	0.69	0.66	0.53	0.91	0.61	0.62
Tomatoes, field grown	0.86	0.79	1.23	1.31	1.43	1.51	1.21

Source: U.S. Bureau of Labor Statistics, CPI Detailed Report, January Issue

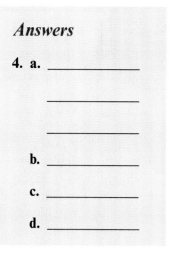

Answers

4. a. _____

_____

_____

b. _____

c. _____

d. _____

Refer to the table above showing retail prices.

4. **a.** Which food items had lower prices in 1996 compared to 1990?

b. What was the percent of change in the price of ice cream from 1991 to 1992? $\left(\textit{Note:}\ \text{Percent of change} = \dfrac{\text{change in price}}{\text{old price} \times 100\%}\right)$

c. What was the percent of change in the price of ice cream from 1990 to 1996?

d. Which food item had the greatest change in price between 1990 and 1996?

Disposable personal income is that amount of money remaining after taxes have been paid. The following table shows disposable personal income per capita by state in current dollars and constant dollars. Constant dollars are current dollars adjusted for inflation.

NO. 707. DISPOSABLE PERSONAL INCOME PER CAPITA IN CURRENT AND CONSTANT (1992) DOLLARS, BY STATE: 1990 AND 1996
[IN DOLLARS. 1996 DATA PRELIMINARY]

REGION, DIVISION, AND STATE	CURRENT DOLLARS		CONSTANT (1992) DOLLARS		REGION, DIVISION, AND STATE	CURRENT DOLLARS		CONSTANT (1992) DOLLARS	
	1990	1996	1990	1996		1990	1996	1990	1996
United States	16,642	20,979	17,912	19,100	District of Columbia .	21,727	29,567	23,385	26,918
					Virginia	17,305	21,434	18,626	19,514
Northeast	**19,254**	**23,995**	**20,723**	**21,845**	West Virginia	12,653	16,494	13,619	15,016
New England	**19,592**	**24,263**	**21,087**	**22,089**	North Carolina	14,568	19,110	15,680	17,398
Maine	15,067	18,219	16,217	16,587	South Carolina	13,644	17,467	14,685	15,902
New Hampshire . . .	18,398	23,329	19,802	21,239	Georgia	15,206	19,664	16,366	17,902
Vermont	15,448	19,381	16,627	17,645	Florida	16,881	21,185	18,169	19,287
Massachusetts	19,806	24,720	21,317	22,505	**East South Central . . . .**	**13,505**	**17,873**	**14,536**	**16,272**
Rhode Island	17,277	21,659	18,595	19,719	Kentucky	13,229	17,192	14,239	15,652
Connecticut	22,715	27,706	24,448	25,224	Tennessee	14,678	19,441	15,798	17,699
Middle Atlantic	**19,135**	**23,901**	**20,595**	**21,760**	Alabama	13,566	17,785	14,601	16,192
New York	19,592	24,380	21,087	22,196	Mississippi	11,578	15,911	12,462	14,486
New Jersey	21,536	26,570	23,179	24,190	**West South Central . . . .**	**14,538**	**18,808**	**15,647**	**17,123**
Pennsylvania . . .	16,880	21,410	18,168	19,492	Arkansas	12,549	16,783	13,507	15,279
Midwest	**16,040**	**20,827**	**17,264**	**18,962**	Louisiana	13,259	17,786	14,271	16,193
East North Central . .	**16,251**	**21,052**	**17,491**	**19,166**	Oklahoma	13,571	16,980	14,607	15,459
Ohio	15,795	20,340	17,000	18,518	Texas	15,307	19,621	16,475	17,863
Indiana	14,970	19,433	16,112	17,692	**West . . . .**	**17,103**	**20,785**	**18,408**	**18,923**
Illinois	17,690	22,778	19,040	20,737	**Mountain . . . .**	**14,724**	**18,753**	**15,848**	**17,073**
Michigan	16,277	21,376	17,519	19,461	Montana	13,140	16,656	14,143	15,164
Wisconsin	15,304	19,858	16,472	18,079	Idaho	13,441	16,722	14,467	15,224
West North Central . .	**15,537**	**20,298**	**16,723**	**18,480**	Wyoming	15,056	18,614	16,205	16,946
Minnesota	16,567	21,597	17,831	19,662	Colorado	16,692	21,265	17,966	19,360
Iowa	14,756	19,723	15,882	17,956	New Mexico	12,898	16,674	13,882	15,180
Missouri	15,461	19,906	16,641	18,123	Arizona	14,562	18,308	15,673	16,668
North Dakota	13,640	18,351	14,681	16,707	Utah	12,395	16,436	13,341	14,964
South Dakota	13,979	19,381	15,046	17,645	Nevada	17,442	21,805	18,773	19,852
Nebraska	15,490	20,180	16,672	18,372	**Pacific . . . .**	**17,933**	**21,557**	**19,301**	**19,626**
Kansas	15,700	20,225	16,898	18,413	Washington	17,179	21,740	18,490	19,792
South	**15,229**	**19,531**	**16,391**	**17,782**	Oregon	15,111	19,189	16,264	17,470
South Atlantic	**16,250**	**20,541**	**17,490**	**18,701**	California	18,315	21,760	19,713	19,811
Delaware	18,591	23,654	20,010	21,535	Alaska	18,100	21,277	19,481	19,371
Maryland	19,151	23,158	20,612	21,083	Hawaii	18,148	21,776	19,533	19,825

Source: U.S. Bureau of Economic Analysis, *Survey of Current Business,* May 1996 and 1997 issues.

5. a. Which state ranked first in disposable personal income per capita in constant dollars in 1990? What is that amount?

b. Which state ranked last in disposable personal income per capita in constant dollars in 1990? What is that amount?

c. What is the dollar difference in per capita disposable personal income in 1990 for these two states?

6. a. Which states ranked first and last in disposable personal income per capita in constant dollars in 1996?

b. What is the dollar difference in 1996 for these two states?

Answers

5. a. _____

_____

b. _____

_____

c. _____

6. a. _____

_____

b. _____

The following table shows per capita spending on libraries in the United States.

FUNDING OF PUBLIC LIBRARIES, 1995

United States	$20.88	Kentucky	$12.22	North Dakota	$12.07
Alabama	12.32	Louisiana	14.79	Ohio	34.68
Alaska	30.12	Maine	18.44	Oklahoma	13.46
Arizona	18.13	Maryland	25.58	Oregon	22.75
Arkansas	9.82	Massachusetts	23.70	Pennsylvania	14.60
California	17.24	Michigan	19.29	Rhode Island	23.60
Colorado	24.00	Minnesota	23.71	South Carolina	12.85
Connecticut	30.83	Mississippi	8.86	South Dakota	18.21
Delaware	14.12	Missouri	18.73	Tennessee	10.18
District of Columbia	34.68	Montana	12.52	Texas	12.06
Florida	17.81	Nebraska	19.78	Utah	18.79
Georgia	14.47	Nevada	17.31	Vermont	16.86
Hawaii	19.94	New Hampshire	20.62	Virginia	21.01
Idaho	17.07	New Jersey	30.99	Washington	28.45
Illinois	30.46	New Mexico	18.26	West Virginia	10.52
Indiana	30.72	New York	36.96	Wisconsin	22.84
Iowa	17.72	North Carolina	13.91	Wyoming	22.88
Kansas	23.80				

Source: National Center for Education Statistics, U.S. Dept. of Education; per capita spending.

Answers

7. a. _____

   b. _____

   c. _____

7. a. Which state spends the most on libraries, per person?

   b. Which state spends the least, per person, on libraries?

   c. What percent of the 50 states spent more on libraries than the national average?

Business Applications 5.3

1. The owner of a chain of fast-food restaurants made the following table of population figures to help decide where in Idaho might be a good place to open her first Idaho store.

City	1990 Population	1994 Population	Percent of Change
Boise	125,738	141,900	_____
Caldwell	18,400	19,129	_____
Coeur d'Alene	24,563	27,200	_____
Idaho Falls	43,929	51,900	_____
Lewiston	28,082	30,400	_____
Moscow	18,519	20,000	_____
Mountain Home	7,913	8,304	_____
Nampa	28,365	33,800	_____
Pocatello	46,080	49,600	_____
Wallace	13,931	13,600	_____

Answers

1. **b.** _____

    c. _____

       _____

    d. _____

       _____

       _____

       _____

       _____

       _____

a. Complete the table. *Note:* Percent of change =

$$\frac{1994 \text{ population} - 1990 \text{ population}}{1990 \text{ population}} \times 100\%$$

b. Which city shows the highest percent of population growth?

c. Based on population alone, which cities might the owner not consider?

d. What other factors might she consider other than population growth when deciding where to locate?

Activities 5.3

Graph Activity

Look at the graph below.

Attitudes about zapping meat

The Agriculture Department soon will announce rules for irradiation of red meat to kill disease microbes, a process approved in December. Food irradiation attitudes by gender and microwave oven use:

	For	Against	No opinion
Men	61%	21%	18%
Women	44%	27%	29%
Microwave often	55%	20%	25%
Microwave rarely/never	40%	36%	24%

Source: Peter D. Hart Research Associates for Grocery Manufacturers of America

By Anne R. Carey and Grant Jerding, USA TODAY

a. Do more men or women favor irradiation? Why do you suppose this is true?

b. If 27,500 people were interviewed, how many men voted For?

c. If 27,500 people were interviewed, how many women voted Against?

d. Out of 27,500 people, how many rarely microwave?

Answers

a. _____

_____

b. _____

c. _____

d. _____

Name:_____   Date:_____

 # Challenge Activity

States get annual federal grants to help pay for drinking water programs. Last year, on average, the federal government paid 41% of states' costs. Here's a look at combined state and federal funding for drinking water programs and the amount spent per household:

State	FY '97 total	Total per household
Alabama	$ 1,133,000	$ 0.70
Alaska	2,655,458	12.41
Arizona	3,126,000	1.85
Arkansas	5,157,844	5.42
California	16,606,020	1.50
Colorado	1,377,230	0.92
Connecticut	N/A	N/A
Delaware	920,952	3.34
Florida	9,899,320	1.75
Georgia	9,670,658	3.55
Hawaii	3,344,838	8.60
Idaho	3,278,482	7.62
Illinois	5,034,900	1.16
Indiana	2,407,333	1.09
Iowa	2,474,427	2.24
Kansas	N/A	N/A
Kentucky	2,685,031	1.82
Louisiana	5,747,168	3.66
Maine	1,298,829	2.69
Maryland	3,827,989	2.05
Massachusetts	4,281,947	1.84

Answers

a._____

b._____

c._____

d._____

e._____

_____

_____

_____

_____

_____

_____

f._____

a. Which state spends the most, in total, on drinking water programs?

b. Which state spends the least, in total, on drinking water programs?

c. Which state spends the most, per household, on drinking water programs?

d. Which state spends the least, per household, on drinking water programs?

e. How do you explain California's position of spending the most, in total, on drinking water problems, to that of Alaska, where they spend $12.41 per household?

f. Approximately how much of the $3,126,000 spent by Arizona is actually paid by that state?

Internet Activity

Have you ever wondered who discovered the planets? Look at the table of Sun, Planet, and Satellite Data and find out the names of the people who discovered Uranus and Pluto. What were the years of the discovery? What else can you learn from this table?

http://bang.lanl.gov/solarsys/eng/data.htm#hi

SKILLBUILDER 5.4

Reading Graphs

Learning Objectives

* **Read and interpret line graphs.**
* **Read and interpret bar graphs.**
* **Read and interpret circle graphs.**

Line Graphs

A **graph** is a visual means of displaying numerical information or data. Graphs are generally less accurate than tables, but they can be used effectively to show trends or comparisons. A **line graph,** or **polygon,** is a form of graph in which information is plotted as points on horizontal and vertical scales. These points are then connected with a broken line. More than one set of data can be shown on a single line graph.

The line graph on the next page shows the sales in one week for three shifts of the clothing department of Dart Round-the-Clock Stores. This graph was constructed by locating, or plotting, the points representing sales for each day (the horizontal scale) on the vertical scale. Each shift was plotted separately, and the points for each shift were then connected. A different type of line was used for each shift.

Student Success Hints

Read for details by going over the key points and main ideas carefully, studying all equations and formulas, and studying the examples and solutions.

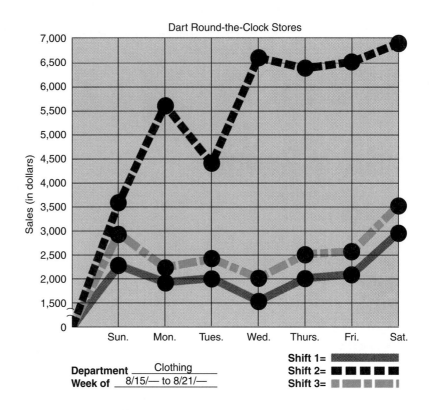

Dart Round-the-Clock Stores

Department ___Clothing___
Week of ___8/15/— to 8/21/—___

Shift 1=
Shift 2=
Shift 3=

Self-Check
Answer
2 days

EXAMPLE

Determine the approximate sales total on Monday for shift 1.

SOLUTION
$2,000

Self-Check
On how many days did shift 3 have sales above $2,500?

Answers

1. a. _____

   b. _____

   c. _____

   d. _____

   e. _____

   f. _____

Problems

Answer the following questions using the information shown on the line graph.

1. **a.** What was the approximate sales total on Wednesday for shift 2?

   b. What was the approximate difference in sales between shift 1 and shift 3 on Saturday?

   c. Which shift had the greatest sales total on Tuesday?

   d. For how many days did shift 2 have sales above $5,000?

   e. Which shift generally has the greatest sales total?

   f. Which shift generally has the least sales total?

Bar Graphs

A **histogram** is a graph that is useful for showing differences between categories. It consists of perpendicular scales, with bars drawn to represent the amounts. The Sportsway Athletic Supply Store histogram shows the results of a survey of the ages of the first 100 customers to make a purchase at the store on a single day. The horizontal scale gives seven age group ranges. The vertical scale gives the number of customers in each age group. The results of the survey are illustrated by the bars drawn on the graph. The top of each bar indicates the number of persons in each age group.

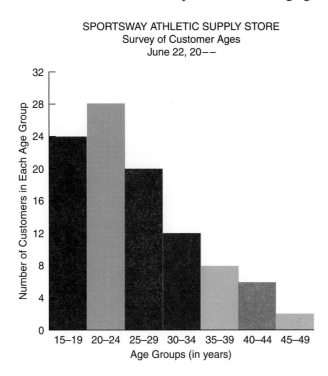

SPORTSWAY ATHLETIC SUPPLY STORE
Survey of Customer Ages
June 22, 20– –

A **bar graph** is similar to a histogram, except that the bars are separated by space. This form of graph is often used to compare data from one period to another. Bar graphs can be drawn either horizontally or vertically and can present data from more than one period. Bar graphs are read in the same manner as histograms. The Dart Round-the-Clock Stores bar graph compares sales this year with sales for last year in the stores' six departments.

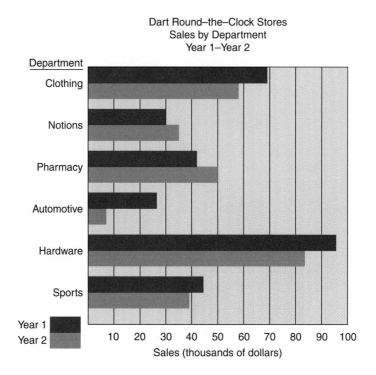

© Glencoe/McGraw-Hill

EXAMPLE

How many customers are in the 30–34-year age group in the Sportsway Athletic Supply Store graph?

SOLUTION

Locate the 30–34-year age group on the horizontal scale. The height of this bar represents the number of customers in this age group. Read across from the top of this bar to the vertical scale on the left. There were 12 customers in this age group.

 Self-Check
According to the Dart Round-the-Clock Stores graph, about how much higher were sales in the Notions Department in year 2 than in year 1?

 **Self-Check Answer**
$3,000

Problems

Use the bar graphs for these problems.

2. Use the histogram to answer the following questions.

   a. Which age group has 20 customers?

   b. How many customers are in the 35–39 year age group?

   c. Which age group had the most customers?

   d. Which age group had the fewest customers?

   e. How many customers were between 30 and 49 years of age?

3. Use the bar graph on page 255 to answer the following questions.

   a. What was the approximate amount of sales in the Notions Department for year 1?

   b. Which departments had sales over $50,000 in year 1?

   c. Approximately how much did automotive sales decrease from year 1 to year 2?

   d. Which departments had higher sales in year 1 than in year 2?

   e. What is the approximate amount of the total sales in year 1 for the Clothing and Sports Departments combined?

Answers

2. a. _____

   b. _____

   c. _____

   d. _____

   e. _____

3. a. _____

   b. _____

   c. _____

   d. _____

   e. _____

Circle Graphs

A **circle graph,** or pie chart, is an effective way to show how individual parts relate to the whole. For example, in the following circle graph, the entire circle represents $1 of federal government income, or 100%. The total income is divided into different categories to indicate the sources of the income: personal income tax, social security tax and Medicare tax (FICA), corporate income tax, and so on. Each of these categories is displayed on a portion of the circle that represents that portion of the income. For example, personal income tax accounts for 42% of every dollar of income.

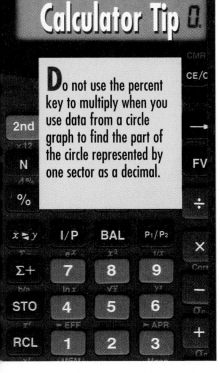

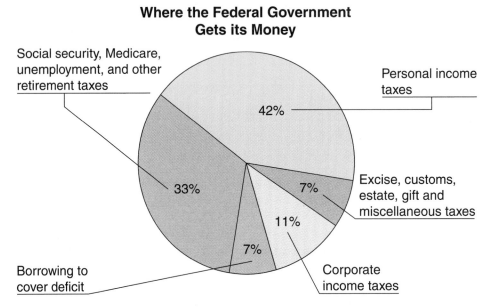

Where the Federal Government Gets its Money

Social security, Medicare, unemployment, and other retirement taxes

Personal income taxes

42%

33%

7%

Excise, customs, estate, gift and miscellaneous taxes

11%

Borrowing to cover deficit

7%

Corporate income taxes

Source: IRS Publ. 11325E

EXAMPLE

How many cents of every dollar come from personal income tax?

SOLUTION
42% comes from personal income tax
42% of $1 = 0.42 × $1.00 = $0.42

Self-Check Answer
18%

Self-Check
What percent of federal income is obtained from borrowing and corporate income tax together?

© Glencoe/McGraw-Hill

Answer the following questions using the circle graph on page 258.

4. a. What percent of every dollar of federal income comes from corporate income tax?

    b. How many cents of every dollar of federal income come from corporate income tax?

    c. What percent of federal income is obtained from social security, medicare, unemployment, and other retirement taxes?

    d. How many cents of every dollar of federal income is borrowed to cover the deficit?

    e. What percent of federal income is obtained from taxes of all types?

    f. What percent accounts for personal income taxes?

    g. What does 11% represent on the graph?

Answers

4. a. _____

   b. _____

   c. _____

   d. _____

   e. _____

   f. _____

   g. _____

5. What percent does the whole graph represent? Why?

6. In what two areas does the federal government get $\frac{3}{4}$ of its money?

7. In what two areas does the federal government get about $\frac{1}{2}$ of its money?

8. What section brings in the most in taxes?

9. What section brings in the least in taxes?

10. What conclusion can you make from this graph?

Business Applications 5.4

Answer the following questions using the following circle graph.

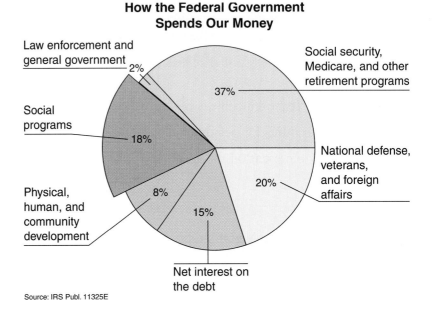

**How the Federal Government
Spends Our Money**

Law enforcement and
general government
2%

Social
programs
18%

Physical,
human, and
community
development
8%

Net interest on
the debt
15%

National defense,
veterans,
and foreign
affairs
20%

Social security,
Medicare, and other
retirement programs
37%

Source: IRS Publ. 11325E

Answers

1. a. _____

   b. _____

   c. _____

   d. _____

1. a. What percent of the federal budget is spent on interest payments?

b. How many cents of every budget dollar are spent on retirement-related programs?

c. What percent of the federal budget is spent on retirement and defense-related programs?

d. What percent of the federal budget is spent on the government's general operating costs and law enforcement?

Name:_____ Date:_____

Answer the questions on page 262 using the following circle graphs.

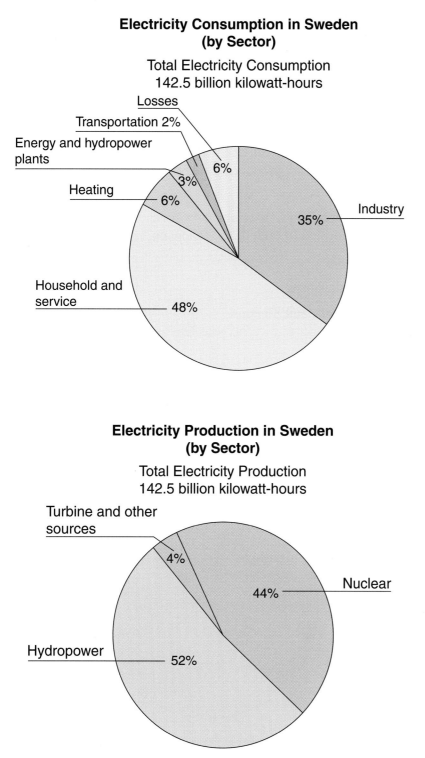

**Electricity Consumption in Sweden
(by Sector)**

Total Electricity Consumption
142.5 billion kilowatt-hours

Losses

Transportation 2%

Energy and hydropower
plants

6%

3%

Heating

6%

Industry

35%

Household and
service

48%

**Electricity Production in Sweden
(by Sector)**

Total Electricity Production
142.5 billion kilowatt-hours

Turbine and other
sources

4%

Nuclear

44%

Hydropower

52%

2. a. Which sector consumes the greatest amount of electricity?

   b. Which sector consumes the least amount of electricity?

   c. What percent of the electricity consumed is used in producing electric power?

   d. Of the total 142.5 billion kilowatt-hours of electricity consumed, how many billion kilowatt-hours of electricity are lost from the system?

   e. Which sector produces the greatest amount of electricity?

   f. Which sector produces the least amount of electricity?

   g. Of the total quantity of electricity produced, how many billion kilowatt-hours are produced using water-powered sources?

3. From the graph below, determine how much of the EPA enforcement budget is allocated to each of the following activities.

   a. Superfund

   b. Hazardous Waste

   c. Lakes, Rivers, and Streams

   d. Air

   e. Toxic Substances

EPA's Enforcement Budget

Total Enforcement Budget $449 million

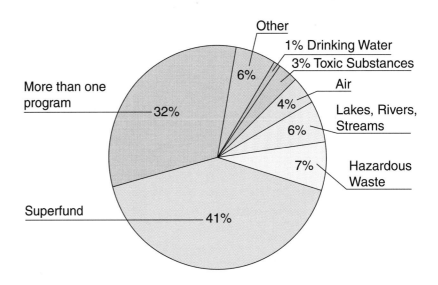

Name:_____ Date:_____

Activities 5.4

Graph Activity

Look at the graph below. What is the total that will be spent on professional books for the three years (1998–2000)? _____

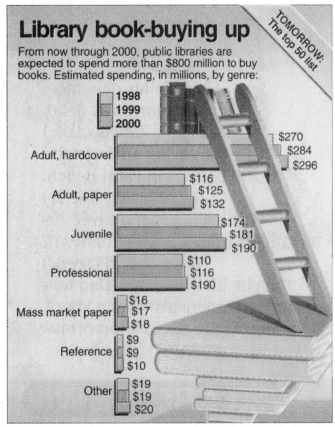

Library book-buying up

From now through 2000, public libraries are expected to spend more than $800 million to buy books. Estimated spending, in millions, by genre:

TOMORROW: The top 50 list

- 1998
- 1999
- 2000

Adult, hardcover: $270 / $284 / $296

Adult, paper: $116 / $125 / $132

Juvenile: $174 / $181 / $190

Professional: $110 / $116 / $190

Mass market paper: $16 / $17 / $18

Reference: $9 / $9 / $10

Other: $19 / $19 / $20

Source: Book Industry Study Group

By Cindy Hall and Suzy Parker, USA TODAY

Challenge Activity

Use the circle graph to answer each question.

Suppose 50 million households carry a credit card balance each month. Round answers to the nearest million.

a. How many households pay their balance in full each month?

b. How many households have no credit cards?

c. If the average outstanding credit card balance each month is $7,000, approximately how much is owed, in total, each month?

Answers

a. _____

b. _____

c. _____

Credit Card Use in the U.S.

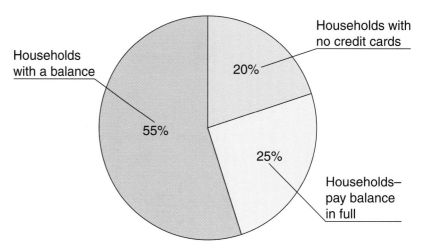

Households with no credit cards

20%

Households with a balance

55%

25%

Households– pay balance in full

Internet Activity

Sometimes it is better to express information in a graph instead of in words. You can find the answers to these questions quickly when you look at the graphs on this site.

http://ecep1.usl.edu/ecep/math/r/r.ht

What is the hardest type of coal? What was the largest source of energy in 1850? What fuel source rose most from 1925 to 1975?

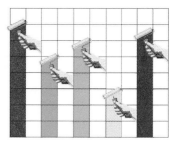

SKILLBUILDER 5.5

Computing Sales Commissions

Learning Objectives

- Compute earnings on a straight commission basis and on a salary plus commission basis.
- Compute earnings on a variable commission basis.

Commissions

Many people who work in sales are paid a percentage of the sales they make. This percentage is called a **commission.** Some people are paid only a commission, and some are paid a commission plus a salary. Others earn a base salary plus a commission only on the amount that exceeds their specified sales quota.

EXAMPLE

Sheila Collins works for Environmental Products and receives a 15% commission on all the sales she makes. Find her commission during a week in which her total sales were $9,800.

SOLUTION

Commission = sales × commission rate

$9,800   Sales
× 0.15   Commission rate
─────────
$1,470   Commission

Sheila's commission is $1,470.

MATH TIP

When figuring sales commission, total sales is the base and the commission rate is the percent.

EXAMPLE

Jamil Toubassi sells furniture for Emmett Designs. He is paid $200 per week plus a 5% commission on all sales. Last week Jamil sold $9,600 worth of furniture. Find his earnings.

SOLUTION

$9,600	Sales		$480	Commission
× 0.05	Commission rate		+ 200	Salary
$480	Commission		$680	Total earnings

Jamil's earnings are $680.

EXAMPLE

Tonya Gibson earns a base salary of $250 per week and an 8% commission on all sales over $2,500. Tonya's sales for the past week were $10,800. Find her earnings for the week.

SOLUTION

$10,800	Sales for week		$8,300	
− 2,500	Minimum quota		× 0.08	Commission rate
$8,300	Sales over $2,500		$664	Commission

$250	Base salary
+ 664	Commission
$914	Total earnings

Tonya's earnings are $914.

Self-Check Answer
$501.20

Self-Check

Emilio Sanchez works at The Outdoor Store. He receives $250 per week plus a commission of 4% on all sales. Find his earnings in a week when his sales were $6,280.

Name:_____ Date:_____

Problems

1. Find the commissions for these salespeople based on their weekly sales.

Salesperson	Rate of Commission	Weekly Sales	Commission
a. Martin Beck	5%	$12,000	_____
b. Jack Longacre	6%	$11,700	_____
c. Jane Hayes	4%	$14,400	_____
d. Alice Bittmore	7%	$10,650	_____
e. Carl Covell	8%	$12,810	_____

2. Find the total of the base salary plus commission for each salesperson based on weekly sales.

Salesperson	Base Salary	Rate of Commission	Weekly Sales	Salary Plus Commission
a. John Bean	$300	5%	$8,250	_____
b. Randy Beasley	$300	6%	6,000	_____
c. Marge Champion	$450	6%	7,600	_____
d. Hilary Dawson	$200	8%	11,100	_____
e. Dawn Gargan	$350	7%	9,750	_____

3. These salespeople receive a base salary of $975 a month and a commission of 6% on all sales over $1,000. Compute the total earnings of each salesperson.

Salesperson	Monthly Sales	Amount of Commission	Total Earnings
a. Mike Dawes	$36,000	_____	_____
b. Pat Hunt	30,320	_____	_____
c. Lois Lipfield	46,800	_____	_____
d. Ruth Moore	85,920	_____	_____
e. Jim Roberts	48,000	_____	_____

Variable Commissions

Some salespeople earn commissions based on a sliding scale. These salespeople may or may not receive a base salary in addition to their commission.

EXAMPLE

Salespeople at Tern Products are compensated according to the following scale:
$0— — $2,000 @ 5% ($2,000 range)
$2,000.01 — $4,000 @ 7% ($2,000 range)
$4,000.01 — Over @ 10%

Celia Chang had sales of $9,200 last week. What were her total earnings for the week?

SOLUTION

Sales × Rate = Commission
$2,000 × 0.05 = $100
2,000 × 0.07 = 140
5,200 × 0.10 = 520
──────   ─────
$9,200 $760
Celia's earnings are $760.

Self-Check Answer
$850

Self-Check
Bryan Tarabochia works at Tern Products, Inc. Find his commission on sales of $10,100.

© Glencoe/McGraw-Hill

Problems

4. Compute the total earnings for each of the following salespeople. The commission paid is based on the sliding scale at Tern Products, Inc.

	Salesperson	Total Sales	Amount of Commission	Base Salary	Total Earnings
a.	Bassett, M.	$ 8,500	_____	$200	_____
b.	Copeland, J.	10,000	_____	None	_____
c.	DelTorro, V.	11,000	_____	230	_____
d.	Englert, T.	9,200	_____	225	_____
e.	Franklin, H.	12,200	_____	230	_____
f.	Fusco, C.	11,600	_____	None	_____
g.	Handley, E.	10,280	_____	225	_____
h.	Montez, W.	10,960	_____	230	_____
i.	Oroz, J.	12,240	_____	225	_____
j.	Pappas, N.	15,200	_____	None	_____
k.	Preston, O.	13,600	_____	225	_____
l.	Restic, V.	8,800	_____	200	_____
m.	Savage, K.	12,400	_____	230	_____
n.	Truax, P.	9,880	_____	200	_____
o.	Weston, A.	12,800	_____	230	_____
p.	Zeramba, L.	6,880	_____	230	_____
	Totals		_____		_____

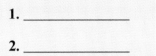

Business Applications 5.5

Answers

1. _____

2. _____

3. _____

4. _____

1. Maria Santangelo is a sales representative for Blitz Lasers. She is compensated for sales (less returns or cancellations) on the following basis:

    Maria's sales for the period were $105,000. A $17,000 order from the previous sales period was cancelled by the customer. How much was her commission for the present period?

2. Pedro Chang is a retail salesperson who is paid an hourly rate of $11, with overtime at time and a half for any hours in excess of 40 per week. In addition, he also receives a $7\frac{1}{2}\%$ commission on all sales over $2,500 per week. Pedro worked a total of 53 hours last week and had sales totaling $8,659. What were his gross earnings for the week?

3. Christine Yen is paid a commission of $8\frac{3}{8}\%$ on all sales, excluding shipping charges. Last month Christine had sales of $12,973, including shipping charges of $324.33; $4,128, including shipping charges of $103.20; $29,697, including shipping charges of $742.43; and $51,097, including shipping charges of $1,277.43. What was her total commission for the month?

4. Amy Snow works at Computers Inc. and earns a 10.5% commission on all software sales. In May, Amy sold $15,704 in software. Compute Amy's commission.

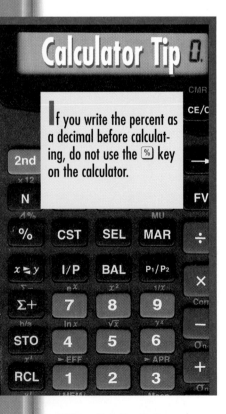

Calculator Tip

If you write the percent as a decimal before calculating, do not use the % key on the calculator.

Activities 5.5

Graph Activity

Ted Ryan works in a sporting goods store in Colorado selling ski equipment and earns a 7.5% commission on all sales. During a recent week Ted sold $2,854 in equipment. Calculate Ted's commission. _____

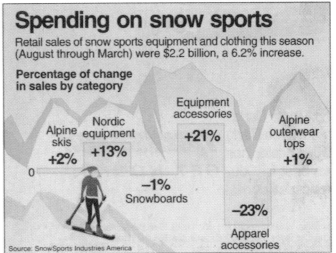

Spending on snow sports

Retail sales of snow sports equipment and clothing this season (August through March) were $2.2 billion, a 6.2% increase.

Percentage of change in sales by category

Alpine skis +2%
Nordic equipment +13%
Equipment accessories +21%
Alpine outerwear tops +1%
−1% Snowboards
−23% Apparel accessories

Source: SnowSports Industries America

By Cindy Hall and Web Bryant, USA TODAY

Challenge Activity

Sam Klarnet has obtained three job offers in sales, and must decide which of the three offers provides the best possibility for earnings. Each of the prospective employers will pay commission on a variable scale. Assuming monthly sales of $50,000 for each of the vendors, and an average work month of 160 hours, which of the following plans should he accept?

a. $1,000 per month plus commission as follows:

$0 to $5,000—0%; $5,001 to $20,000—5%; $20,001 to $40,000—$7\frac{1}{2}$%; $40,001 and over—10%.

b. Commission only: $0 to $5,000—5%; $5,001 to $20,000—$7\frac{1}{2}$%; $20,001 and over—10%.

c. $6 per hour plus commission: $0 to $2,500—0%; $2,501 to $15,000—5%; $15,001 to $40,000—$7\frac{1}{2}$%; $40,001 and over—10%.

Internet Activity

How are commissions paid in the DTR Software International Company?

http://www.dtr-software.com

If you need additional practice in working with percents go to

http://www.testprcp.com/

Name:_____ Date:_____

SKILLBUILDER 5.6

Computing Agents' and Brokers' Commissions

Learning Objective

• **Compute earnings for agents and brokers.**

Agent and Broker Commissions

Agents and brokers are individuals who work for themselves but who represent the business interests of others. They are usually compensated on a commission basis. A sales representative who is employed by a company may be paid a salary plus a commission and is usually reimbursed for expenses incurred in selling. Agents or brokers may represent manufacturers, importers, exporters, athletes, artists, or property owners. The contract between the representative and the principal (client) determines the basis for compensation.

EXAMPLE

Peter Pfaffenroth is an automobile broker who receives $7\frac{1}{2}\%$ commission on all sales he concludes. How much commission will he receive on the sale of a 1927 Ford for $18,500?

SOLUTION

$Commission = Sales \times Rate\ of\ Commission$
$\$1,387.50 = \$18,500 \times 0.075$

Peter's commission is $1,387.50.

Self-Check
Pavi Terhar is a real estate broker. She receives a 7% commission on the sale of a house. How much commission does she receive on a house that sells for $198,000?

> **MATH TIP**
>
> Remember, a misplaced decimal point will cause an error in the amount of commission.

Self-Check Answer
$13,860

Problems

Langford Boutique pays its representatives $6.50 per hour plus a commission of 2.5% of all sales. Compute the earnings for each representative.

Name	Hours Worked	Sales	Earnings
1. Goldsmith, A.	23	$1,320	_____
2. Palusinska, K.	40	$2,544	_____
3. Trinh, P.	32	$2,065	_____
4. Varga, J.	33	$1,955	_____
5. Zurlinden, M.	40	$2,200	_____

Answers

6. _____

7. _____

8. _____

9. _____

Solve each problem.

6. Arlene Jablonski had sales of $250,000 for the months of January through March. She is paid a commission of 8% on all sales. What was her commission for the quarter?

7. Pedro Chang is paid 10% commission on sales to retailers and 6% commission on sales to wholesalers. If sales to retailers were $90,000 and sales to wholesalers were $175,000, what was the total commission he earned?

8. Nicole Koshak is paid 12% commission on net sales (sales minus returns). Sales for the period were $50,000, with returns of $1,800. What was her commission?

9. Harold Huizenga is paid $9.50 an hour plus 3% of all sales. How much did he earn in a week in which he worked 32 hours and sold $6,300?

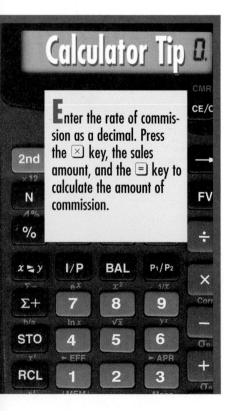

Calculator Tip

Enter the rate of commission as a decimal. Press the ⊠ key, the sales amount, and the ⊟ key to calculate the amount of commission.

© Glencoe/McGraw-Hill

Business Applications 5.6

1. David Hurt sells pianos at Music World. He is paid a commission based on his monthly sales. For the first $10,000 in sales, he is paid 8% and for sales exceeding $10,000, he is paid 15%. David sold $25,500 worth of pianos during December. Find the amount of David's gross pay from commissions for December.

2. Samantha Ho sells office furniture at Professional Office, Inc. She is paid a commission based on her monthly sales. For the first $5,000 in sales, she is paid 5% and for sales exceeding $5,000, she is paid 10%. Samantha sold $21,800 worth of furniture during September. Find the amount of Samantha's gross pay from commissions for September.

3. Arwen Hill is the manager of a restaurant. She is paid a salary of $500 per month plus 3% of the net income of the restaurant. In October, the restaurant's net income was $59,000. Find Arwen's gross pay for the month of October.

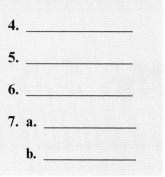

Answers

4. _____

5. _____

6. _____

7. a. _____

   b. _____

4. CheJan Park, an importer's broker, sold a total of $185,000 of imported manufactured products to various importers. CheJan receives a commission a $4\frac{3}{4}\%$ on all orders sold. How much did she earn on this group of sales?

5. Bernadette Pavley, an artist's agent, receives 7% commission on all concert fees paid her artists, 3% commission on all records sold at public appearances, and $1\frac{1}{2}\%$ commission on all product endorsements. Total concert fees for the year were $3,679,000; record sales totaled $273,000; product endorsement contracts totaled $1,936,852. What was Bernadette's gross income for the year?

6. Marion Castello, a real estate sales agent, sold a commercial property for $875,000. Her broker receives an 8% commission on the sale, and Marion receives 30% of the total commission. How much is Marion's commission on this sale?

7. Noel Parisi is an importer of gourmet food products, which he imports on specific order. (a) Compute the total invoice value on an order of 30 cases of tuna at $23 per case, 16 cases of mushrooms at $13.95 per case, 25 cases of Mandarin oranges at $15 per case, 20 cases of crabmeat at $34.50 per case, and 5 cases of Dijon mustard at $19.80 per case. In addition there is a shipping charge of $1.58 per case, and an import document charge of $25. (b.) If Noel's commission is $2.15 per case, how much was his commission?

276   Unit 5   Percents

© Glencoe/McGraw-Hill

Name:_____ Date:_____

Activities 5.6

Graph Activity

Rob Riley, an agent who represents several pro football players, receives a $3\frac{1}{2}\%$ commission on all the players' fees for television appearances to endorse products. Last year the players Rob represented earned $4,720,000 in fees for television appearances. What was Rob's commission? _____

Familiar grounds

The Cowboys and 49ers are the only teams reporting to new training sites this summer. Longest active NFL training camp tenures in years:

Team, training camp
Packers, St. Norbert College
Vikings, Mankato State Univ.
Steelers, St. Vincent College   32
N.Y. Jets, Hofstra University   30

40
32

Source: NFL   By Scott Boeck and Gary Visgaitis, USA TODAY

Challenge Activity

Rhea Germain brokers retail and commercial leasing of passenger cars, commercial vehicles, and office equipment. The commission earned is 15% on all retail leasing, 12% on commercial leases, and 8% on leases of office equipment. For the period just ended, she completed auto and truck leases totalling $893,000, of which 78% was related to commercial leasing of cars and trucks. Office equipment leases totalled $128,000. Compute Rhea's commission earnings for this period. _____

Internet Activity

Find out the sales commission of U.S. Pride Realty Co. For what service is this low commission increased? What is the increase?

http://www.pubcenter.com/uspride/u.s.p

Read more about who pays commission for real estate transactions at

http://homeadvisor.msn.com/highlight/qa/5254.as

SKILLBUILDER 5.7
Computing Markup and Selling Price

Learning Objectives

- **Compute markup and percent of markup based on cost price.**
- **Compute markup and percent of markup based on selling price.**
- **Find the markup and selling price.**

Markup Based on Cost Price

Retailers make a profit and cover their expenses by charging more for a product than they pay for it. The amount added to the cost price is the **markup.** The markup rate is called the **percent of markup,** which can be based on the cost price or on the selling price.

Most retail merchants use the rate of markup based on cost to determine their selling price. For example, if a retail merchant decides that all items must have a 25% markup based on cost, an item costing $24 would be sold for $30.

Cost × Rate of Markup Based on Cost Price = Markup
$24 × 0.25   = $6
Cost + Markup = Selling Price
$24 + $6   = $30

Another way to look at the problem above is cost **always** equals 100% and selling price will **always** be more than 100%. So you can think of finding cost (C), markup (M), rate of markup based on cost price (R), or selling price (S) by adding 100% to the rate of markup based on cost.

$C \times (100\% + R) = S$
$24 \times 125\%$ $= S$
24×1.25 $= \$30$

If the same merchant decides to use a 50% markup based on cost, the item costing $24 would sell for $36.

$24 \times 0.50 = \$12$ $24 + \$12 = \$36$

Student Success Hints

You can think of the problem at the left like this.

100%	C	$24
25%	+ M	6
125%	S	$30

MATH TIP

Remember to enter the decimal in the selling price and to change the markup percent to a decimal.

Student Success Hints

You can think of the problem at the right like this.

100%	C	$37.50
$16\frac{2}{3}\%$	+ M	6.25
$116\frac{2}{3}\%$	S	$43.75

Self-Check Answers
a. $110
b. 50%

EXAMPLE

An item costs $37.50; its selling price is $43.75. Find (a) the markup and (b) the rate of markup based on the cost price.

SOLUTION

a.
$43.75	Selling price	(S)
− 37.50	Cost price	(C)
$6.25	Markup	(M)

b. *Rate of Markup Based on Cost Price = Markup ÷ Cost Price*

$$16\tfrac{2}{3}\% \quad \text{or} \quad 16.67\% \;=\; \$6.25 \;\div\; \$37.50$$

Self-Check

A chair costs $220; its selling price is $330. Find (a) the markup and (b) the rate of markup based on the cost price.

If a merchant decides to use a 20% rate of markup based on cost, and the item sells for $18, what is the cost?

$$S \div R = C$$
$$\$18 \div 120\% = C$$
$$\$18 \div 1.20 = \$15$$

The item costs $15.

Student Success Hints

You can think of the problem at the right like this.

100%	C	?
20%	+ M	?
120%	S	$18

Suppose a sweater costs $32 and a merchant marks it up $24, what is the rate of markup based on cost?

$$M \div C = R$$
$$\$24 \div \$32 = 75\%$$

The sweater has been marked up 75%.

Student Success Hints

You can think of the problem at the right like this.

100%	C	$32
?%	+ M	24
?%	S	$56

© Glencoe/McGraw-Hill

Problems

Solve each problem.

1. Julie Carter purchased radios for $18 each and sold them for $24.

   a. Find the markup.

   b. Find the rate of markup, to the nearest hundredth of a percent, based on cost price.

2. Pierre Products purchased electronic typewriters for $150 each and sold them for $191.60.

   a. Find the markup.

   b. Find the rate of markup, to the nearest hundredth of a percent, based on cost price.

3. Caryl Clothes purchased a dozen blouses at $216 a dozen. Each blouse retailed for $23.31. What was the rate of markup, to the nearest tenth of a percent, based on cost price?

4. Goliath Corp. purchased a gross of calculators for $3,730. (There are 144 items in a gross.) The calculators will be priced at a markup of $37\frac{1}{2}$ percent of the cost price. What will be the selling price of each calculator?

Student Success Hints

Remember to use the five-step problem-solving plan on page 57 to solve problems. The steps include reading the problem, deciding what information is needed, choosing the operation, solving the problem, and checking the answer.

5. The Alberta Shoe Emporium realizes a profit of $13 on each pair of a certain style of shoe that it sells. This represents a 50 percent markup based on the cost price.

   a. What is the cost price of each pair of shoes?

   b. What is the selling price of each pair?

Markup Based on Selling Price

Markup can also be based on the selling price, or retail price, of the goods. Selling price will **always** equal 100%. Cost % and Markup % **each** are **always** less than 100%.

Self-Check Answer
30%

EXAMPLE

Find the rate of markup based on the selling price for a bicycle that cost $75 and that sells for $87.50.

SOLUTION

$87.50	Selling price
− 75.00	Cost price
$12.50	Markup

Rate of Markup
Based on Selling Price = Markup ÷ Selling Price
14.29% = $12.50 ÷ $87.50

Self-Check
A TV costs $201.60. The markup is $86.40. Find the percent of markup based on the selling price.

Answers

6. a. _____

   b. _____

7. a. _____

   b. _____

Problems

Solve each problem.

6. Portable typewriters cost a dealer $144 each. The markup is $66.

    a. What is the selling price of each typewriter?

    b. What is the rate of markup to the nearest hundredth of a percent based on the selling price?

7. A living room chair cost $270 and sells for $324.

    a. What is the amount of the markup?

    b. What is the rate of markup to the nearest hundredth of a percent based on the selling price?

Finding Markup and Selling Price

Cost price divided by cost price expressed as a percent of selling price will yield the selling price.

75%	C	$37.50
25%	+ M	?
100%	S	?

EXAMPLE

A radio cost $37.50 and will be sold at a markup of 25% based on the selling price. Find the markup and the selling price.

SOLUTION

Selling price	100%	
Markup	− 25%	of selling price
Cost Price	75%	of selling price

Selling Price = Cost Price ÷ Percent Cost Price Is of Selling Price

$50.00 =	$37.50 ÷	75%

$50.00   Selling price
− 37.50   Cost price
$12.50   Markup

Self-Check

A calculator cost $48 and will be sold at a markup of $37\frac{1}{2}$% based on the selling price. Find (a) the markup and (b) the selling price.

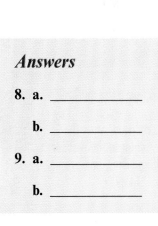

Self-Check Answers
a. $28.80
b. $76.80

Problems

Solve each problem.

8. A suitcase cost $43.56 and was sold at a markup of 50% of the selling price.

   a. What was the selling price?

   b. What was the amount of markup?

9. A digital alarm clock cost $18 and was sold at a markup of 40% of the selling price.

   a. What was the selling price?

   b. What was the amount of markup?

Answers

8. a. _____

   b. _____

9. a. _____

   b. _____

Business Applications 5.7

1. Complete this table. Round off the rate to the nearest hundredth of a percent, where necessary. Round to the nearest cent where necessary.

	Cost Price	Markup	Selling Price	Rate of Markup Based On	
				Cost Price	Selling Price
a.	$86.40	_____	$103.68	_____	_____
b.	46.20	6.60	_____	_____	_____
c.	_____	2.48	_____	16%	_____
d.	_____	4.80	_____	_____	8.75%

Answers

2. _____

3. a. _____

   b. _____

   c. _____

4. _____

2. Fancy Frocks spring line includes designer shorts, which cost them $17.38. Their usual markup is 70% of selling price. What is the selling price?

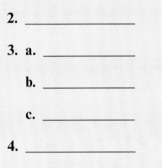

3. Home-Pro Tools sell overstock tools from their outlet stores at an average markup of $37\frac{3}{8}\%$ of cost. What should the selling price be for each of the following?

   a. A cordless screwdriver costing $11.85.

   b. A $\frac{3}{8}$-inch drill costing $21.50.

   c. $7\frac{1}{4}$-inch circular saw costing $31.38.

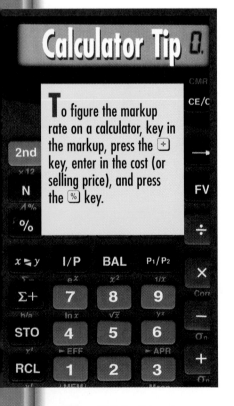

4. The Jewelry Chest manufactures custom jewelry to order, which they usually mark up at 250% of production cost. The cost for design, labor, and materials for a gold pendant was $87.50. What was their selling price for this item?

Activities 5.7

Graph Activity

a. A bookstore purchased 500 copies of *The Notebook* at a cost of $3.10 each. If the bookstore marks this book up by 40% over cost, what will be the selling price?

b. For every 1.4 copies of *Sugar Busters!* sold, how many copies did *Dr. Atkins' New Diet Revolution* sell?

c. Of both hardcover and paperback, which book sold the most?

d. Which book of all sells more?

Answers

a. _____

b. _____

c. _____

d. _____

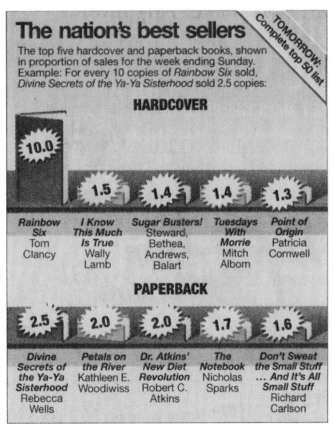

Challenge Activity

Mortar & Pestle determined that the cost of research and development of a new antibiotic is equal to 78 percent of their production cost for the product. In addition they require another 35 percent to cover overhead, marketing costs, and profit. (Carry to 4 decimal places.)

a. With a production cost of 95 cents per capsule, what price must the manufacturer charge the wholesaler for this product?

b. The wholesaler requires a 24 percent markup on his cost to cover all operating costs and profit. What price must the wholesaler charge the retail pharmacist?

c. The retail pharmacist marks up 20 percent on the wholesaler's selling price. What price must the pharmacist charge the consumer for this antibiotic?

Answers

a. _____

b. _____

c. _____

Internet Activity

Use a table to find the markup. If you want to earn a 40% profit, how much would you charge for an item that costs $4.00? How much would you charge to make a 60% profit?

http://www.smartbiz.com/sbs/arts/sbs77.ht

Binders $4.00

SKILLBUILDER 5.8

Computing Markdown and Selling at a Loss

Learning Objectives

* Compute the amount and rate of markdown; the reduced, or new, selling price; and the original selling price.
* Compute the cost price and selling price when goods are sold at a loss.

Computing Markdown

A reduction in selling price, called **markdown,** is used to move slow-selling merchandise. Markdown is expressed either as an amount or as a rate (percent). The rate of markdown may be based on either the original selling price or the new, or reduced, selling price. The dollar differences are the same, but using the reduced selling price gives the impression of a larger markdown.

EXAMPLE

An item that sells for $10 is marked down to $8. What are the amount of markdown and the rates of markdown based on the original selling price and on the new selling price?

SOLUTION

$10 Original selling price
− 8 New selling price
$ 2 Markdown

Rate of Markdown = Markdown ÷ New Selling Price
      0.25 = 25% = $2   ÷ $8

Rate of Markdown = Markdown ÷ Original Selling Price
      0.20 = 20% = $2   ÷ $10

The markdown is $2. The rate of markdown based on the new selling price is 25%. The rate based on the original selling price is 20%.

Self-Check
An item that sells for $25 is marked down to $18. Find the rate of markdown based on the original selling price.

MATH TIP

To find the new selling price, find the complement of the markdown rate. The complement is found by subtracting the discount from 100%

Self-Check Answer
28%

Problems

Solve the following problems.

1. Find the markdown for the merchandise below. Find the rate of mark-down to the nearest tenth of a percent, on both original selling price and new selling price.

	Original Selling Price	New Selling Price	Markdown	Rate of Markdown	
				Original Price	New Price
a.	$25.00	$20.00	_____	_____	_____
b.	84.00	70.00	_____	_____	_____
c.	99.00	88.00	_____	_____	_____
d.	527.00	395.25	_____	_____	_____

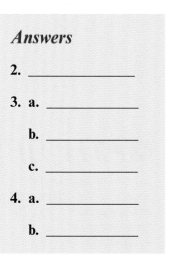

Answers

2. _____

3. a. _____

   b. _____

   c. _____

4. a. _____

   b. _____

2. The Varsity Shop marked down golf clubs from $94.80 to $79. What was the rate of markdown based on the new selling price?

3. Billy Bootery marked down shoes from $50 to $40.

   a. What was the rate of markdown based on the new selling price?

   b. If the shoes were marked down an additional $15, what was the total rate of markdown based on the new selling price?

   c. What was the total rate of markdown based on the original selling price?

4. A bicycle is marked down $30, which is 20% of the original selling price.

   a. What was the original selling price?

   b. What is the new selling price?

Selling at a Loss

Goods are sometimes sold at a loss. When this is done, the new selling price is less than the cost of the goods. When the percent of loss is based on the new selling price, the cost price of the item is always equal to 100 percent plus the percent of loss based on the new selling price.

Student Success Hints

Because mathematics builds one step at a time, it is important to complete and understand your homework.

EXAMPLE

a. Find the cost price of an item which was sold for $3.60 at a loss of $16\frac{2}{3}$ percent based on the selling price.

b. Find the new selling price of an item on which the loss is $12.32, which is $33\frac{1}{3}\%$ of the new selling price.

SOLUTION

a.
$$\begin{array}{ll} 100\% & \text{Selling price} \\ + \ 16\frac{2}{3}\% & \text{Loss} \\ \hline 116\frac{2}{3}\% \text{ or } 116.67\% & \text{Cost price} \end{array}$$

Cost Price	=	New Selling Price	×	Percent Cost Price Is of Selling Price
	=	$3.60	×	$116\frac{2}{3}\%$
$4.200 or $4.20	=	$3.60	×	1.1667

b.

New Selling Price	=	Amount of Loss	÷	Loss as a Percent of New Selling Price
$36.96	=	$12.32	÷	$33\frac{1}{3}\%$
Cost	=	New Selling Price	+	Amount of Loss
$49.28	=	$36.96	+	$12.32

Self-Check
Find the new selling price of an item that cost $24.64 if it is sold at a loss of $33\frac{1}{3}$ percent based on the new selling price.

Self-Check Answer
$18.48

Calculator Tip

If you are given the rate of markdown, you can use the percent discount key to find the net price.

Solve the following problems.

5. Complete the following table. Round off the rate of loss to the nearest tenth of a percent, where necessary.

	Cost Price	New Selling Price	Loss	Rate of Loss Based On	
				Cost Price	New Selling Price
a.	$106.80	$83.20	_____	_____	_____
b.	2.10	_____	0.42	_____	_____
c.	6.50	_____	_____	_____	20%
d.	_____	13.65	_____	_____	12.5%
e.	11.20	8.96	_____	20%	_____

Answers

6. a. _____

   b. _____

7. a. _____

   b. _____

6. A sewing machine was sold for $72 less than its cost price, which was $37\frac{1}{2}\%$ of the new selling price.

   a. What is the new selling price?

   b. What did the sewing machine cost?

7. An overstock 2 head VCR, with a list price of $149, and normally sold at $90, is clearance priced at $13 less than its cost price. The loss on the sale was $16\frac{2}{3}$ percent of the cost price.

   a. Find the cost price.

   b. What is the new selling price?

Name:_____   Date:_____

Business Applications 5.8

Answers

1. _____

2. _____

3. _____

4. _____

1. Green Village Gardens is selling a group of decorative pots at $3.20, which is a 25% loss. Their usual markup is 40% of selling price. What was the original selling price of the pots?

2. McCarter Wholesalers is overstocked on kitchen gadget sets normally selling for $25. On this class of merchandise, McCarter has a markup of 15% on cost. To clear the inventory, it is offering the sets at a loss of 15%. What should the new selling price be?

3. The Elegant Boutique has marked down designer jeans three separate times, to a final price of $18. The markdowns were $33\frac{1}{3}$%, 15%, and 25%. What was the original selling price?

4. Ferguson's Books overstocked *In The Kitchen With Rosie,* and needs to lower the price. The book normally sells for $14.95, but Ferguson's marks it down 30%. What is the new selling price?

5. Finlandia Importers are overstocked on leaded crystal bowls, normally selling at $23.75. Their actual cost is $15.25. In order to clear the inventory, they are planning a series of three separate markdowns, 15 percent, 25 percent, and 40 percent. What would each of these markdown prices be? Would any of these markdown prices be below cost?

6. Van's Appliances paid $600 for a big screen TV. Their regular markup is 60% on selling price. The TV did not sell at the regular price, was marked down by 30%, and later by another 20%. The TV still did not sell, so Van marked the price up by 10%, and advertised it as a special promotion. What was the original selling price, the markdown prices, and the final price?

Activities 5.8

Graph Activity

Nature's Harvest Health Food Store sells St. John's Wort in bottles of 100 for $15.99. The store had a sale on this product and marked it down by 40%. What was the price of St. John's Wort during this sale? Round your answer to the nearest cent. _____

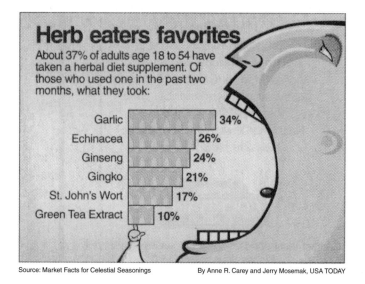

Herb eaters favorites

About 37% of adults age 18 to 54 have taken a herbal diet supplement. Of those who used one in the past two months, what they took:

Garlic	34%
Echinacea	26%
Ginseng	24%
Gingko	21%
St. John's Wort	17%
Green Tea Extract	10%

Source: Market Facts for Celestial Seasonings   By Anne R. Carey and Jerry Mosemak, USA TODAY

Answers

a. _____
_____
_____

b. _____
_____
_____
_____

c. _____
_____
_____

d. _____
_____
_____

Challenge Activity

Dippy Doug's Appliance and Electronics store has declared bankruptcy, and you have an opportunity to buy from its remaining inventory below cost. For each of the following items, given the dollar amount of the loss, percent of loss on the final selling price, and rate of markup, find the final selling price, actual cost, and the original list price for:

a. A 32-inch TV marked up 25% on selling price, and sold at $25 below cost, which is a 5% loss on the final selling price.

b. A Home Theater unit marked up 40% on selling price, and sold at $80 below cost, which is a 20% loss on final selling price.

c. A 21 cubic foot refrigerator marked up 70% on cost, and sold at $100 below cost, which is a 10% loss on final selling price.

d. A dishwasher marked up 35% on selling price, and sold at $25 below cost, which is an $8\frac{1}{3}\%$ loss on final selling price.

Internet Activity

Check out the Student Discount Network. Choose Boulder, Colorado and go to the Colorado Bookstore. If you print this coupon what would be the percent of discount on a $25 purchase? How long is the coupon valid?

http://www.discount-net.com/cu.ht

SKILLBUILDER 5.9

Using Ratios and Proportions

Learning Objectives

- **Determine if two ratios are proportional.**
- **Solve proportions for the missing value.**

Ratios and Proportions

A **ratio** is the relationship of one number to another. A ratio may be written either by separating the numbers by the word *to*, by separating the numbers by a colon (:), or by showing the relationship as a fraction.

For example, if Gardner Electronics employs 30 full-time production workers and 10 part-time production workers, the ratio of full-time to part-time workers is

$$30 \text{ to } 10 \quad or \quad 30:10 \quad or \quad \frac{30}{10}$$

This ratio is generally written in simplest form as $3:1$. When the fraction form is used, only two numbers can be compared, but when a colon or the word *to* is used, more than two numbers can be compared. For example, if Gardner Electronics also employs five office and clerical workers, the ratio of full-time production workers to part-time production workers to office employees is

$$30 \text{ to } 10 \text{ to } 5 \quad or \quad 30:10:5 \quad or \quad 6:2:1$$

A **proportion** is a comparison of two sets of numbers having the same ratio or relationship. In the previous example, the ratio of office workers to total employees is 5 to $(30 + 10 + 5)$, or $5:45$. In fraction form, this ratio can be reduced to $\frac{1}{9}$. These two ratios, when compared, form a proportion, as follows:

$$\frac{5}{45} = \frac{1}{9}$$

This expression is read "5 is to 45 as 1 is to 9."

Student Success Hints

The more practice you complete for each skillbuilder, the firmer a foundation you are building for more complex mathematics. Ask your instructor for practice beyond the homework if you feel you need it.

Since the ratio, or relationship, is the same between the two sets of numbers, their cross products will be equal. In cross multiplication, the numerator of each ratio is multiplied by the denominator of the other ratio. The resulting products must be equal for the ratios to be a proportion.

$$\frac{5}{45} \boxtimes \frac{1}{9} \qquad 5 \times 9 = 45$$
$$1 \times 45 = 45$$

The two ratios are in proportion.

MATH TIP

Expressions such as 7 out of 8 can be written as 7/8.

Self-Check
Answer
Yes

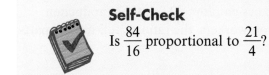

EXAMPLE

Is $\frac{2}{3}$ proportional to $\frac{4}{9}$?

SOLUTION

$$\frac{2}{3} \boxtimes \frac{4}{9} \qquad 2 \times 9 = 18$$
$$3 \times 4 = 12$$

Since the cross products are not equal, the ratios are not proportional.

Self-Check
Is $\frac{84}{16}$ proportional to $\frac{21}{4}$?

Name: _____ Date: _____

Problems

Determine whether the given ratios are proportional.

1. $\frac{5}{9}, \frac{20}{36}$

2. $1:1\frac{1}{2}, 3:4\frac{1}{2}$

3. $\frac{5}{12}, \frac{7}{15}$

4. $\frac{3}{8}, \frac{9}{24}$

5. $1:2, 60:180$

Calculator Tip

A ratio suggests division. When solving proportions, use the ⊠ and ⊞ keys.

Solving Proportions

Where two sets of numbers are proportional, but one set is incomplete, the unknown value can be found by cross multiplying.

EXAMPLE

Gardner Electronics is planning an increase in its production line. Three wrapping machines require an area of 28 m^2. How much space must Gardner provide for a total of 12 wrapping machines?

SOLUTION

Known
Number of machines presently in use—3
Number of machines to be used—12
Area required for 3 machines—28 m^2

Unknown
Area required for 12 machines

Ratios
The ratios involved are 12 machines to 3 machines and N m^2 to 28 m^2. The proportion would be 12 is to 3 as N is to 28.

1. $\dfrac{12}{3} = \dfrac{N}{28}$

2. Cross multiply 12 by 28: $12 \times 28 = 336$.

3. Place this product over the denominator 3, and carry out the division.

$$N = \frac{336}{3} = 112 \text{ m}^2$$

4. **Check**

$\dfrac{12}{3} = \dfrac{112}{28}$ $12 \times 28 = 336$
                $3 \times 112 = 336$

Twelve machines would require 112 m^2 of floor space.

Self-Check Answer
400

Self-Check
Solve for N: $\dfrac{3}{75} = \dfrac{16}{N}$.

Problems

Solve for the unknown in the following:

6. $\dfrac{1}{N} = \dfrac{5}{15}$

7. $\dfrac{10.5 \text{ gal}}{250 \text{ mi}} = \dfrac{G \text{ gal}}{300 \text{ mi}}$

8. $\dfrac{18}{23} = \dfrac{36}{N}$

9. $\dfrac{N}{17} = \dfrac{96}{102}$

10. $\dfrac{49}{7} = \dfrac{X}{1}$

11. $\dfrac{2}{Y} = \dfrac{10\frac{1}{2}}{420}$

12. $\dfrac{H \text{ h}}{25 \text{ mi}} = \dfrac{3 \text{ h}}{5 \text{ mi}}$

13. $18 : 3 = 126 : X$

14. $\dfrac{4}{56} = \dfrac{C}{784}$

15. $8 : 0.77 = 32 : N$

Business Applications 5.9

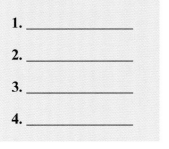
1. At Gilray Electronics, it takes 16 production workers to produce 48 parts in 1 h. How many workers would be needed to produce 144 parts in 1 h?

2. Fred's Flowers has a new delivery van, which used 25 L of gasoline to travel 290 km. How far can the van travel on 60 L of gasoline?

3. Hamburger Heaven can prepare 136 portions from 34 lb of ground beef. How many portions can be prepared from 110 lb of ground beef?

4. At Tycee Manufacturing, a quality check found 21 transformers defective out of a production run of 840. Inspection of a second lot found 67 defective units. Assuming a uniform defect rate, how many transformers were in the second lot?

Activities 5.9

 ## Graph Activity

Look at the graph below. It took 120 people surveyed between 10 and 19 minutes to drive to work. How many people surveyed took 30 to 44 minutes? Round to the nearest whole number. _____

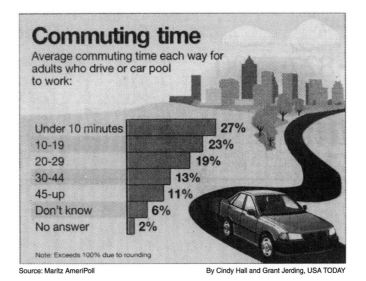

Commuting time

Average commuting time each way for adults who drive or car pool to work:

Under 10 minutes	27%
10-19	23%
20-29	19%
30-44	13%
45-up	11%
Don't know	6%
No answer	2%

Note: Exceeds 100% due to rounding

Source: Maritz AmeriPoll   By Cindy Hall and Grant Jerding, USA TODAY

 ## Challenge Activity

For one week, determine the ratio of hours spent in school and studying to free time. Write ratios for other activities such as hours spent at a job, hours spent sleeping, hours working around the house, all activities combined, etc., to actual free time. Once you have described a typical week in ratios, then write them for a day, a month, and a year.

 ## Internet Activity

If the ratio of boys to girls is 5 to 8, what do you need to know to find out the number of boys and girls in a class?

http://www.lerc.nasa.goiv/Other_Groups/k-12/p_test/math/9mathe_p_q10.ht

UNIT 6 BUSINESS AND CONSUMER MATH

In this unit, you will study the following Skillbuilders:

Banking is changing rapidly. Consolidations and mergers are resulting in fewer, but larger banks. Banking consumers of all types are still interested in such factors as location, friendliness, variety and cost of services, and interest rates, but are now increasingly looking for electronic banking features, such as ATMs and Internet banking by reducing, or eliminating, personal contact between the customer and the bank.

As human contact becomes more remote, it is essential that every consumer of banking services have a working knowledge of the way in which banking transactions take place, thus, in this unit we discuss checking accounts and statements, payroll plans, inventory, and depreciation.

Nation's Leading Banks Ranked by Assets in Billions of Dollars

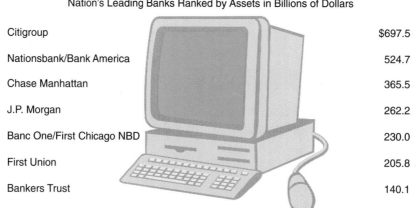

Citigroup	$697.5
Nationsbank/Bank America	524.7
Chase Manhattan	365.5
J.P. Morgan	262.2
Banc One/First Chicago NBD	230.0
First Union	205.8
Bankers Trust	140.1

MATH CONNECTIONS

Accountant

Judy Jones, a tax accountant, uses her training to gather pertinent information, analyze data, and make recommendations to her clients, based on her knowledge and her understanding of the clients' desires. Judy will then present her findings and recommendations in a readable, understandable format so that her clients can make the best choice. Often, tax accountants like Judy recommend that clients open individual retirement accounts (IRAs) as a reasonable way to prepare for retirement and defer federal income tax.

Math Application

Judy's clients, Bob and Jean Holmes, are concerned about retirement but have little extra money to invest. After some discussion, Judy recommends that the Holmes deposit $150 per month in a tax-deferred IRA through Jean's employer. Bob wants to know if this means that Jean will be lowering her take-home pay by $150 each month. Assume the Holmes would normally pay a tax rate of 20% on the $150.
Compute the saved taxes.

$$(20\%)\,(150) = \underline{\hspace{2cm}}$$

Subtract the saved taxes from the $150 to compute the net pay difference.

$$\$150 - \$30 = \underline{\hspace{2cm}}$$

Critical Thinking Problem

The Holmes insist that even $50 a month lower income will cause hardship. What might Judy suggest, to help lower the burden of the IRA?

SKILLBUILDER 6.1

Checking Accounts and Bank Records

Learning Objective

- **Complete a check register or a check stub.**

Student Success Hints

More is better. Solve extra problems. You can also change the numbers in problems you have already completed and do them again.

Checking Accounts

A checking account is essential for a business. It allows the business to pay its employees and other bills, as well as to maintain an accurate account of its income and spending.

A **deposit slip** is used to deposit money to an account. The deposit slip shown here indicates that Brenda's Towing Service deposited two checks in the amounts of $85.88 and $195.

DEPOSIT TICKET

Brenda's Towing Service
Route 22
Kelsey, OR 97140

DATE _____Nov. 7_____ 20 _____
DEPOSITS MAY NOT BE AVAILABLE FOR IMMEDIATE WITHDRAWAL

_____
SIGN HERE FOR CASH RECIEVED (IF REQUIRED)

Security National Bank
Kelsey, OR 97140

⊕⑈:123202280⑈:   15 056 0⑈'   12

Cash	Currency		
	Coin		
LIST CHECKS SINGLY 92–36		85	88
1–118		195	00
Total From Other Side			
TOTAL		280	88
Less Cash Received		—	
NET DEPOSIT		280	88

96-228/ 1232

Use other side for additional listing.

Be certain each item is prpoerly endorsed.

Checks and other items are recieved for deposit subject to the provisions of the Uniform Commercial Code or applicable collection agreement.

A check is written on the account to pay a bill, for example. When a check is presented to a bank, the bank is authorized to take the amount specified from the account on which the check is drawn. In the check shown here, Brenda's Towing Service has paid $99.58 to Pacific Power for its October electric bill.

Date check written

Preprinted check number

Brenda's Towing Service
Route 22
Kelsey, OR 97140

1228

November 8 20 _____ 53-393/113

Code number to identify bank

To whom check is payable, or payee

PAY TO THE ORDER OF *Pacific Power* $ *99 58/100*

Amount of check

Verbal form of amount of check. Note spacing and use of "and" to represent the decimal.

Ninety-nine and 58/100 _____ DOLLARS

Security National Bank
Kelsey, OR 97140

MEMO *October bill* *Brenda Moore*

Signature

⑆ 123202280 ⑆ 15 056 0 ⑈ 1228 0000009958

Bank number printed with magnetic ink for computer processing matches printed number at upper right-hand corner of check above amount of check

Brenda's Towing Service account number

Preprinted check number

When bank processes check, the 99.58 is imprinted here. Note that this should match what is written for amount of check.

To keep track of deposits made, checks written, and the balance on hand, the depositor uses a **checkbook stub** or **check register.** The stub or register entries should always be filled out before the check is written.

CHECK NO.	DATE	CHECKS ISSUED TO OR DESCRIPTION OF DEPOSIT	AMOUNT OF PAYMENT	✓	AMOUNT OF DEPOSIT	BALANCE FORWARD	
						2,348	56
—	11/7	To *Deposit*			280 88	Check or Dep. 280	88
		For				Bal. 2,629	44
1228	11/8	To *Pacific Power*	99 58			Check or Dep. 99	58
		For *October electric service*				Bal. 2,529	86
—	11/9	To *Bank service charge*	10 00			Check or Dep. 10	00
		For				Bal. 2,519	86
		To				Check or Dep.	
		For				Bal.	
		To				Check or Dep.	
		For				Bal.	
		To				Check or Dep.	
		For				Bal.	
		To				Check or Dep.	
		For				Bal.	

Whenever a deposit is made, the amount is added to the balance shown in the register.

The amount of any check is deducted from the balance.

Any charges the bank makes for its services should be deducted, and any money collected for the depositor should be added.

It is important to properly identify each transaction and to add or subtract as the nature of the transaction requires.

© Glencoe/McGraw-Hill

Name:_____ Date:_____

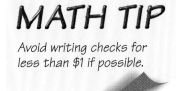

EXAMPLE

Find the balance in the checking account on the previous page if a check in the amount of $215.90 is written to Cost Club.

MATH TIP

Avoid writing checks for less than $1 if possible.

SOLUTION

The amount of a check is subtracted:

$2,519.86
− 215.90
$2,303.96   Balance

Self-Check

Find the balance in this checking account after checks totaling $213.25 are deposited and a check in the amount of $199.52 is written.

Self-Check Answer

$2,317.69

Problems

Solve these problems.

1. This check register shows various checks written and deposits made. The balance before check number 576 is $485.14. Assuming all entries are correctly written, what is the balance remaining after check number 581 is written?

		RECORD ALL CHARGES OR CREDITS THAT AFFECT YOUR ACCOUNT					BALANCE FORWARD	
NO.	DATE	CHECKS ISSUED TO OR DESCRIPTION OF DEPOSIT	AMOUNT OF PAYMENT	✓	OTHER DEDUCT	AMOUNT OF DEPOSIT	485	14
576	6/2	To Connally-Wright For Insurance	112 75					
577	6/3	To MasterCard For May statement	116 86					
	6/5	To Deposit For ———				175 00		
578	6/6	To United Compaign For Contribution	10 00					
579	6/8	To Public Gas & Electric Co. For May Gas & Electric	75 00					
	6/12	To Deposit For ———				300 00		
580	6/15	To State Motor Vehicles For License renewal	19 50					
	6/15	To Printing charge For New check supply	14 31					
581	6/18	To West Penn Oil Co. For Fill tank	343 20					
		To For						

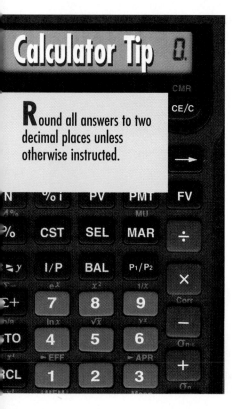

Calculator Tip

Round all answers to two decimal places unless otherwise instructed.

Business Applications 6.1

1. The following six check stubs show the amount of each check and deposits made. The balance before check number 535 is $1,240.78. What is the balance after check number 540 is written? _____

		DOLLARS	CENTS
NO. _535_	$ _10.50_		
DATE _January 29, 20--_			
TO _City Clerk_			
FOR _Permit_			
BALANCE			
AMT. DEPOSITED		695	96
TOTAL			
AMT. THIS CHECK			
BALANCE			

		DOLLARS	CENTS
NO. _536_	$ _69.84_		
DATE _January 29, 20--_			
TO _Perry Press_			
FOR _Advertising flyers_			
BALANCE			
AMT. DEPOSITED			
TOTAL			
AMT. THIS CHECK			
BALANCE			

		DOLLARS	CENTS
NO. _537_	$ _33.45_		
DATE _January 30, 20--_			
TO _Postal Service_			
FOR _Postage_			
BALANCE			
AMT. DEPOSITED			
TOTAL			
AMT. THIS CHECK			
BALANCE			

		DOLLARS	CENTS
NO. _538_	$ _580.16_		
DATE _January 30, 20--_			
TO _Champion Sales_			
FOR _Shirts-Invoice 873 A_			
BALANCE		326	59
AMT. DEPOSITED		246	74
TOTAL			
AMT. THIS CHECK			
BALANCE			

		DOLLARS	CENTS
NO. _539_	$ _118.91_		
DATE _February 8, 20--_			
TO _Headley Paper Co._			
FOR _Wrapping supplies_			
BALANCE			
AMT. DEPOSITED			
TOTAL			
AMT. THIS CHECK			
BALANCE			

		DOLLARS	CENTS
NO. _540_	$ _481.05_		
DATE _February 18, 20--_			
TO _City Tax Collector_			
FOR _First quarter tax_			
BALANCE			
AMT. DEPOSITED			
TOTAL			
AMT. THIS CHECK			
BALANCE			

Activities 6.1

Graph Activities

Tom's mother took him school shopping. She wrote a check for $38.12 for school supplies, $69.90 for sneakers, and $185.40 for clothes. Until these checks have cleared the bank, assuming everything else is in order, how much higher is the bank balance compared to the checkbook balance?
_____

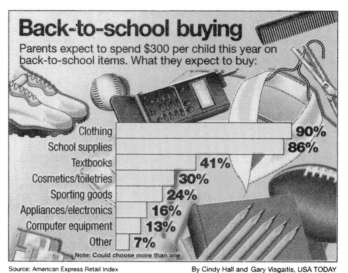

Back-to-school buying

Parents expect to spend $300 per child this year on back-to-school items. What they expect to buy:

Clothing	**90%**
School supplies	**86%**
Textbooks	**41%**
Cosmetics/toiletries	**30%**
Sporting goods	**24%**
Appliances/electronics	**16%**
Computer equipment	**13%**
Other	**7%**

Note: Could choose more than one

Source: American Express Retail Index   By Cindy Hall and Gary Visgaitis, USA TODAY

Challenge Activity

Big Pier Imports is preparing to deposit the store receipts for the day. The receipts include $3,829 in currency; $1,183.61 in coin; various checks totalling $9,709.83; and a total of $13,921.17 in charge slip purchases, for which the bank charges a fee of $2\frac{1}{2}$ percent, deducted in advance.

a. Find the net amount of this deposit. _____

b. Add this deposit to the previous balance of $4,987.29. _____

Internet Activity

Express checking accounts which are designed for customers who prefer to bank by ATM, telephone, or personal computer are being offered by the top 35 metropolitan markets in the country. These accounts offer unlimited check writing, low minimum balance requirements and low—or no—monthly fees. What are the disadvantages to these accounts?

wysiwyg://20/http://www.bankrate.com/brm/news/chk/19980608.as

SKILLBUILDER 6.2

Reconciling Bank Statements

Learning Objective

- **Reconcile a bank statement**

Bank Statements

Every month the bank sends a statement for each checking account, summarizing all transactions that have taken place since the previous statement. The bank statement should promptly be compared to the checkbook record. Adjusting the checkbook balance and the bank statement balance is called **reconciliation.** The differences in the balances that may appear can be due to outstanding checks written but not yet cleared by the bank, deposits in transit (deposits made too late to be included on the statement), service charges, credits, or interest earned on the account.

MATH TIP

If there is an error in the checkbook balance, add or subtract the amount of the error at the end of the register rather than changing each amount.

EXAMPLE

A typical bank statement is shown below. The form printed on the back, which is used to assist in reconciling the account, is shown on page 313. The bank statement shows a balance of $515.67. The balance on the last checkbook entry is $528.95. Use the following steps in reconciling the balances.

FIRST NATIONAL BANK OF SECAUCUS
Secaucus, NJ 07094

Reginald Perry
532 Sesame Lane
Secaucus, NJ 07094

ON	YOUR BALANCE WAS	NO.	WE SUBTRACTED CHECKS TOTALLING	LESS ACTIVITY CHARGE	NO.	WE ADDED DEPOSITS OF	MAKING YOUR PRESENT BALANCE
Sep. 10, 20--	561.71	15	1220.77	0.00	4	1174.73	515.67

DATE	CHECKS			DEPOSITS	BALANCE
Sep. 10					561.71
Sep. 11	27.19				534.52
Sep. 16				318.53	853.05
Sep. 18	4.00	22.20			826.85
Sep. 21	15.00				811.85
Sep. 22	25.00			126.59(T)	913.44
Sep. 24	124.45	5.18(S)			783.81
Sep. 25	24.05				759.76
Sep. 30				225.00	984.76
Oct. 1	5.10	10.00	29.40	500.53	1440.79
Oct. 6	443.80				996.99
Oct. 7	18.99	100.00			878.00
Oct. 8	101.84	264.57			511.59
Oct. 10				4.08(I)	515.67

SOLUTION

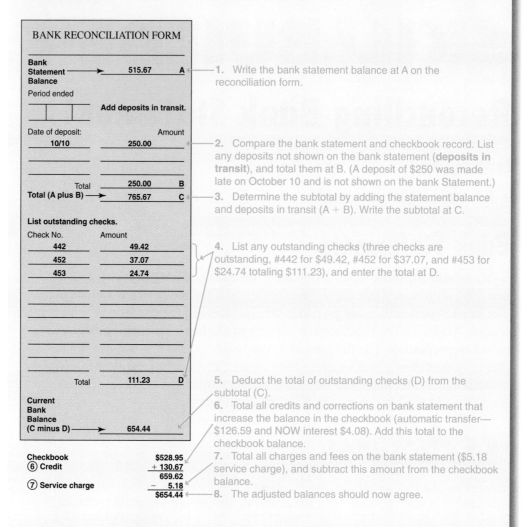

BANK RECONCILIATION FORM

Bank Statement Balance	515.67	A

Period ended

Add deposits in transit.

Date of deposit:	Amount	
10/10	250.00	
Total	250.00	B
Total (A plus B) →	765.67	C

List outstanding checks.

Check No.	Amount	
442	49.42	
452	37.07	
453	24.74	
Total	111.23	D

Current Bank Balance (C minus D) →	654.44	

Checkbook	$528.95	
⑥ Credit	+ 130.67	
	659.62	
⑦ Service charge	− 5.18	
	$654.44	

1. Write the bank statement balance at A on the reconciliation form.

2. Compare the bank statement and checkbook record. List any deposits not shown on the bank statement (**deposits in transit**), and total them at B. (A deposit of $250 was made late on October 10 and is not shown on the bank Statement.)

3. Determine the subtotal by adding the statement balance and deposits in transit (A + B). Write the subtotal at C.

4. List any outstanding checks (three checks are outstanding, #442 for $49.42, #452 for $37.07, and #453 for $24.74 totaling $111.23), and enter the total at D.

5. Deduct the total of outstanding checks (D) from the subtotal (C).

6. Total all credits and corrections on bank statement that increase the balance in the checkbook (automatic transfer—$126.59 and NOW interest $4.08). Add this total to the checkbook balance.

7. Total all charges and fees on the bank statement ($5.18 service charge), and subtract this amount from the checkbook balance.

8. The adjusted balances should now agree.

Self-Check

The Happy Dog, a pet store, had a bank statement balance of $2,367.94. Deposits in transit totalled $1,619.88. Outstanding checks totalled $2,817.42. The account earned $10.58 in interest, and the checkbook balance was $1,149.82. Do the adjusted balances agree?

Self-Check Answer

no

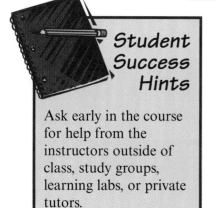

Student Success Hints

Ask early in the course for help from the instructors outside of class, study groups, learning labs, or private tutors.

Problems

Reconcile each checkbook balance with the bank balance. Give the new checkbook balance.

1. Checkbook balance: $1,938.03   Bank balance: $1,796.43
   Unreturned checks: $260.10; $400; $10.98; $18.50; $2
   Interest: $7.94
   Deposits not shown: $841.12

2. Checkbook balance: $218.08   Bank balance: $463.85
   Unreturned checks: $14.52; $126.78; $43.92; $165.80
   Service Charge: $5
   Deposits not shown: $100.25

3. Checkbook balance: $1,632.95   Bank balance: $1,099.38
   Unreturned checks: $15; $7.21; $6; $18
   Service charge: $2
   Deposits not shown: $577.78

4. Barry Goldman started a pet-walking service. At the end of the first month, his checkbook showed a balance of $546.92. When he received his bank statement, he found that he had been charged $18.75 for personalized checks, $5 service charge, and $25 for overdrawing his account during the month. What should his bank balance actually be?

Answers

1._____

2._____

3._____

4._____

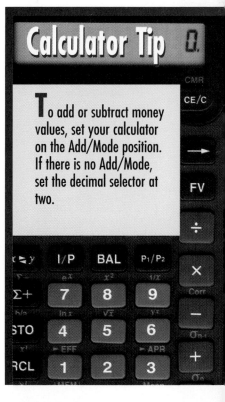

Calculator Tip

To add or subtract money values, set your calculator on the Add/Mode position. If there is no Add/Mode, set the decimal selector at two.

Business Applications 6.2

Complete the following bank reconciliations.

1. Karl's Lawnmower Service received a bank statement showing a balance of $605.50, from which a service charge of $7.50 had been deducted. Karl's Lawnmower Service's checkbook showed a balance of $448.16. Three checks were outstanding, totalling $164.84.

2. Dauber Dairies' bank statement showed a balance of $1,755.97. Two notes for $300 and $515 had been collected by the bank that month and credited to the account. There was a collection charge of $7.50 and a service charge of $4.50. The checkbook balance was $765.53. Outstanding checks totalled $187.44.

BANK RECONCILIATION FORM

Bank Statement Balance ——————→ _____ A
Period ended

| | | Add deposits in transit. |

Date of deposit: Amount

Total _____ B
Total (A plus B) ——→ _____ C

List outstanding checks.

Check No.   Amount

Total _____ D

Current Bank Balance (C minus D) ——→ _____

BANK RECONCILIATION FORM

Bank Statement Balance ——————→ _____ A
Period ended

| | | Add deposits in transit. |

Date of deposit: Amount

Total _____ B
Total (A plus B) ——→ _____ C

List outstanding checks.

Check No.   Amount

Total _____ D

Current Bank Balance (C minus D) ——→ _____

3. Julia's Catering Service's checkbook showed a balance of $224.03. The bank statement indicated a balance of $312.45, which included a service charge of $5.42. Two checks were outstanding totalling $309, and Julia's deposit of $215.16, dated April 25, was in transit.

4. The Corliss Corporation received its monthly bank statement showing a balance of $4,055.16. A $50 check, which had been deposited, was returned for insufficient funds. The bank also collected a $125 note for Corliss, for which they charged $3.50. The checkbook balance was $4,026.36, with outstanding checks totalling $157.30. There was a deposit in transit in the amount of $200 dated January 30.

BANK RECONCILIATION FORM

Bank Statement Balance ———▶ _____ A

Period ended

| | | | **Add deposits in transit.** |

Date of deposit: Amount

_____   _____
_____   _____
_____   _____

Total _____ B
Total (A plus B) ———▶ _____ C

List outstanding checks.

Check No.   Amount

_____   _____
_____   _____
_____   _____
_____   _____
_____   _____
_____   _____
_____   _____
_____   _____

Total _____ D

Current Bank Balance (C minus D) ———▶ _____

BANK RECONCILIATION FORM

Bank Statement Balance ———▶ _____ A

Period ended

| | | | **Add deposits in transit.** |

Date of deposit: Amount

_____   _____
_____   _____
_____   _____

Total _____ B
Total (A plus B) ———▶ _____ C

List outstanding checks.

Check No.   Amount

_____   _____
_____   _____
_____   _____
_____   _____
_____   _____
_____   _____
_____   _____
_____   _____

Total _____ D

Current Bank Balance (C minus D) ———▶ _____

Activities 6.2

Graph Activity

If the number of Internet users trading stocks on line increases by 250% from the fourth quarter of 1997 to the first quarter of 2000, how many users will then be trading stocks on line? _____

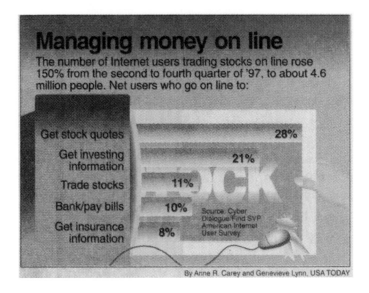

Managing money on line

The number of Internet users trading stocks on line rose 150% from the second to fourth quarter of '97, to about 4.6 million people. Net users who go on line to:

Get stock quotes — 28%
Get investing information — 21%
Trade stocks — 11%
Bank/pay bills — 10%
Get insurance information — 8%

Source: Cyber Dialogue Find SVP American Internet User Survey

By Anne R. Carey and Genevieve Lynn, USA TODAY

Challenge Activity

Despina Dukasis' bank statement shows a balance of $1,852.23. Her checkbook balance was $1,990.81. A check in the amount of $244.14 was outstanding. A deposited check in the amount of $135.60 was returned unpaid, for which the bank charged $20. Three ATM withdrawals, totalling $225, showed on the bank statement. There was a $3.50 charge for one of these. Interest earned for the period was $1.38. Prepare a reconciliation.

_____

Internet Activity

Bank mergers have caused a fierce pricing war. What are the best checking account deals in the United States? Find out what the cheapest check accounts offer.

wysiwyg://15/http://www.bankrate
.com/brm/news/chk/19981201.as

Name:_____ Date:_____

SKILLBUILDER 6.3

Payroll Plans

Learning Objectives

- **Compute gross earnings on an hourly wage basis.**
- **Compute gross overtime earnings.**
- **Compute gross earnings on a piecework basis.**

Hourly Rate Payroll Plan

Many employees are paid a fixed hourly rate of pay, or a **regular hourly wage rate. Total (gross) earnings** are calculated by multiplying the hours worked by the wage for each hour worked.

EXAMPLE

Nannette Moreno worked the following hours one week: 8, 7, 8, 7, and 8. At a rate of \$9.80 an hour, how much did she earn that week?

SOLUTION

$8 + 7 + 8 + 7 + 8 = 38$ h

$$
\begin{array}{rl}
\$9.80 & \text{Hourly rate} \\
\times\ 38 & \text{Hours} \\
\hline
\$372.40 & \text{Gross earnings}
\end{array}
$$

Self-Check
Ken Sandovan worked 8, $7\frac{1}{2}$, $8\frac{1}{2}$, 8, and 6 h one week. At a rate of \$7.75 per hour, how much did he earn that week?

Self-Check Answer
\$294.50

Problems

Complete the following partial payroll register to find each employee's **gross earnings.** The employee's actual take-home pay will be less than the gross earnings once deductions are made for social security, income taxes, and so on.

1.

ISLER MANUFACTURING COMPANY WEEK ENDING MAY 18, 20—								
Employee Number	**Hours Worked**					**Regular Hours**	**Hourly Rate**	**Gross Earnings**
	M	**T**	**W**	**TH**	**F**			
375	8	8	8	8	8	_____	$ 8.00	_____
411	8	8	8	8	4	_____	9.25	_____
465	8	$7\frac{1}{2}$	8	6	8	_____	11.50	_____
505	4	4	8	8	8	_____	8.45	_____
573	8	8	7	8	8	_____	7.25	_____
							Total	_____

Overtime

Employees receive their regular wage rate for time worked within the normal workweek, which is usually 40 hours. Time worked beyond the normal 40 hours is called **overtime.**

Federal law provides that employees shall be paid at a rate at least 50% greater than their regular hourly rate for all hours in excess of 40 per week. This is called **time and a half.** For example, at time and a half, a normal hourly rate of $9.80 would give an overtime rate of $14.70 an hour (9.80 × 1.5 = $14.70). When employees work on Sundays or holidays, they sometimes receive overtime rates that are double their regular hourly rate.

MATH TIP

You can write time and a half as 1.5 or as $1\frac{1}{2} = \frac{3}{2}$ to simplify computation.

EXAMPLE

During one week, Dee Morrison worked 45 h. If her regular hourly rate is $8.50, what were her gross earnings for the week if she is paid time and a half for all hours worked over 40?

SOLUTION

Her overtime rate is	$8.50 × 1.5	= $ 12.75
Regular earnings	40 h × $8.50 =	340.00
Overtime earnings	5 h × $12.75 =	63.75
Total gross earnings		$403.75

Name:_____ Date:_____

Self-Check

Roberto Sanchez worked $46\frac{1}{2}$ h one week. If his regular hourly rate is $6.88, what were his gross earnings for the week if he is paid time and a half for all hours over 40?

Self-Check Answer
$342.28

Find the salary in each case if overtime (hours over 40) is paid time and a half.

2. 46 h; $6.50 per hour

3. 39 h; $9.55 per hour

4. 42 h; $8.48 per hour

5. 45 h; $7.98 per hour

6. Andrea earns $11.15 per hour and double time if she works on Sunday. How much did she earn if she worked 39 h, 4 of which were worked on Sunday?

Answers

2._____

3._____

4._____

5._____

6._____

Piecework Payroll Plan

Manufacturing employees are sometimes paid according to the number of items they complete. This is the **piecework wage plan.** The rate paid may vary with the difficulty of the task or the experience of the employee.

Student Success Hints

Studying for tests.
Review all text, notes, handouts, and other materials in the week before the test.

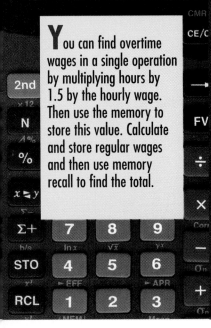

EXAMPLE

During a 5-day workweek, Marv Grasso completed 74 items. If the rate for each piece was $2.775, what were Marv's gross earnings?

SOLUTION

74 × $2.775 = $205.35

Self-Check
Betty Wilson completed 99 items and was paid $2.775 for each item. Find her gross earnings.

Self-Check Answer
$274.73

Problems

7. A partial payroll register for Singleton Instrument Company is shown below. Determine the total number of articles made by each employee. Determine gross earnings by multiplying total articles times rate. Round off gross earnings to the nearest cent.

SINGLETON INSTRUMENT COMPANY PAYROLL REGISTER

Employee Number	Number of Articles					Total Articles	Rate	Gross Earnings
	M	T	W	TH	F			
51	64	62	65	61	65	_____	$2.00	_____
52	64	65	60	62	—	_____	2.20	_____
53	68	65	66	65	62	_____	2.10	_____
54	59	62	60	64	66	_____	2.25	_____
55	62	65	63	59	56	_____	$2.455	_____
56	58	—	55	52	60	_____	2.50	_____
57	64	68	69	64	65	_____	2.25	_____
58	65	61	66	66	64	_____	2.775	_____
							Total	

Business Applications 6.3

1. The payroll register for Rob's Appliance Repair is shown here.
Employees work overtime when the company has more repair projects
to do than can be completed during a normal working day. Complete
the payroll.

ROB'S APPLIANCE REPAIR
WEEK ENDING OCT. 24, 20—

Employee Number	Hours Worked					Time		Hourly Rate	Wage		Gross Earnings
	M	T	W	TH	F	Regular	Overtime		Regular	Overtime	
110	8	8	8	8	8	_____	_____	$8.50	_____	_____	_____
111	8	8	8	9	4	_____	_____	9.75	_____	_____	_____
112	8	9	9	8	8	_____	_____	11.00	_____	_____	_____
113	9	10	8	9	8	_____	_____	10.20	_____	_____	_____
114	3	8	9	9	10	_____	_____	14.00	_____	_____	_____
115	10	9	9	9	9	_____	_____	12.35	_____	_____	_____
116	8	8	8	8	8	_____	_____	7.50	_____	_____	_____
117	—	10	10	9	9	_____	_____	7.10	_____	_____	_____
118	8	9	$8\frac{1}{2}$	8	8	_____	_____	8.25	_____	_____	_____
119	9	9	9	9	7	_____	_____	15.00	_____	_____	_____
								Totals	_____	_____	_____

2. Millie Gonyea's regular pay rate is $14. Find her overtime pay rate if her employer pays time and a half for overtime.

3. Tim Randall earns $8.10 an hour. For the week ending January 9, he worked 30 hours. Calculate his gross pay for the period.

4. Carmen Raines works for a company that makes decorative bottles. She is paid $0.10 per unit for fitting corks to the bottles coming off the assembly line. Carmen fitted 3,400 bottles with corks this week. Find her gross pay for the week.

5. Paul Mattel works in retail sales, and is compensated on an hourly rate plus commission. He is paid $5.35 an hour for the first 40 hours; time and one-half for all hours over 40, except for Sundays and holidays, when his hourly rate is doubled. In addition, he receives a 2 percent commission on all sales. Compute Paul's earnings for a week in which his total sales were $17,826, and he worked the following hours: Sunday—6 hours; Tuesday—8; Wednesday—11; Thursday—9; Friday—9; Saturday—6 hours. (Note—Monday was observed as Veteran's Day.)

Activities 6.3

Graph Activity

Moriah Central School currently employs 56 people. Of these employees, 30 are teachers, 6 are administrators, and 20 are staff workers. Moriah Central School plans to increase the number of teachers by 10%, the number of administrators by 50%, and the number of staff workers by 15%. After these people are hired, how many employees will be working at Moriah Central School? _____

Help wanted this summer

Workforce increases are planned this summer by 32% of companies, the best quarterly outlook since the third quarter of 1978, when 34% were hiring. Companies planning increases, by industry:

Construction	40%
Wholesale/retail	36%
Durable goods manufacturing	33%
Services	31%
Non-durable manufacturing	28%
Finance/insurance/real estate	28%
Education	27%
Transport/utilities	26%
Public administration	25%

Source: Employment Outlook Survey, Manpower   By Cindy Hall and Sam Ward, USA TODAY

Challenge Activity

Tri-Star Production pays its assembly line workers on the following variable piece rate basis:

1 to	125 units @	$0.625	each
126 to	500 units @	0.6875	each
501 to	900 units @	0.75	each
901 to	1,200 units @	0.85	each
1,201 up	units @	$0.95	each

Compute the gross earnings for these employees:

a. McLaren—1,311 units _____   **b.** Brendan—695 units _____

c. Choi—1,526 units _____

Internet Activity

The number of consumers who are experimenting with online banking is growing rapidly as the number of people with home computers increases. Consumers are searching for convenience and value. ATM machines, telephone banking, or banking by mail are often unnecessary. Find out the services available through electronic banking.

http://www.bankrate.com   Select Online banking.

SKILLBUILDER 6.4

Payroll Deductions

Learning Objective

- **Use appropriate tables to compute payroll deductions.**

Payroll Deductions

A **payroll register** for Lund's Tool and Die is shown on page 328. There are four columns in the Employee Data section. The Number column is the number assigned to each employee by the employer. This number is commonly the individual's social security number. The remaining columns show the employee's name, marital status, and the number of exemptions claimed for tax withholding purposes.

The Earnings section contains four columns. The Hours column shows the total number of hours the employee worked during the pay period. The Regular column shows the employee's regular earnings—40 or fewer hours worked times the hourly rate. The Overtime column shows any overtime earned. The Total column shows the employee's total gross earnings—regular wages plus any overtime.

The Deductions section contains seven columns. FIT is the federal income tax deducted from the employee's total earnings. This deduction is determined from tables provided by the Internal Revenue Service (IRS) based on the payroll period, marital status, and the number of exemptions claimed. Both federal and state income taxes are mandatory taxes collected on all taxable earnings without limit. Examples of these wage bracket tax tables are shown in the Appendix.

Social security tax (SS Tax) is another mandatory tax, collected at a flat rate, without reference to marital status or exemptions. At this time, SS Tax is collected at a rate of 6.2% on the first $72,600 of gross earnings. Medicare tax (Med Tax) is collected on a similar basis at a rate of 1.45% on all gross earnings. The employer is required to match both of these amounts. Most states and some cities have instituted a personal income tax, the exact basis being determined by the individual state or city. Union dues are commonly deducted from members' paychecks. AAP is a voluntary program whereby employees who choose to participate contribute 1% of their gross earnings to fund a local social program.

The Amount column in the Net Pay section shows the employee's actual take-home pay, which is total earnings minus total deductions. The employee paycheck number is recorded in the Ck. No. column.

Student Success Hints

To study for tests—prepare summary sheets listing all major concepts, equations, and formulas.

MATH TIP

You can find the social security and medicare taxes by multiplying gross earnings by the rates of 6.2% and 1.45% respectively.

EXAMPLE

Sam Purdy is married and claims three exemptions. His gross earnings for the week are $671. How much federal income tax, social security tax, and medicare tax are withheld from his weekly earnings?

SOLUTION

Select the income tax table for married persons located in the Appendix. In the left-hand column, locate the wage bracket that contains Sam's gross earning, $671 ("At least $670 but less than $680"). Move across the table to the column headed "3." The tax to be withheld is $59. Next to find the social security tax, multiply $671 × 6.2% (671 × 0.062). The social security tax is $41.60. To find the medicare tax multiply $671 × 1.45% (671 × 0.0145). The medicare tax is $9.73.

Self-Check Answer
$499.48

Self-Check

LuAnn Riggin is married and claims four exemptions. Her gross earnings for the week are $582. Find her net pay after federal income tax, social security tax, and medicare tax are deducted.

Problems

Using the tables provided in the Appendix, compute deductions and net pay for each of the following problems.

	Gross Earnings	Exemp.	Marital Status	Federal Tax	SS Tax	Med Tax	Total	Net Pay
1.	$469.90	1	M	_____	_____	_____	_____	_____
2.	710.00	5	M	_____	_____	_____	_____	_____
3.	500.01	8	S	_____	_____	_____	_____	_____
4.	611.75	0	M	_____	_____	_____	_____	_____
5.	578.20	3	S	_____	_____	_____	_____	_____
6.	411.80	2	S	_____	_____	_____	_____	_____

Business Applications 6.4

Answers

1. a. _____

   b. _____

   c. _____

   d. _____

2. _____

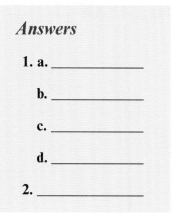

1. Jim Allen's gross earnings last week were $489.99. He is single and claims two exemptions. How much (a) income tax, (b) social security tax, and (c) medicare tax would be deducted from his pay? (d) What is his net pay?

2. Helen Baque worked 43 hours last week. Her hourly rate is $12.50, with overtime paid at a rate of time-and-one-half. Helen is married and claims four exemptions. What is her net pay?

3. Complete the payroll register for Lund's Tool and Die, shown on page 328, using the tables provided in the Appendix. State tax is computed at 3% of gross earnings. For those employees who participate, 1% of their gross earnings is deducted for AAP. Check numbers begin with number 1411. Total all money columns.

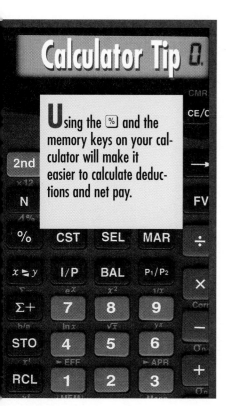

Calculator Tip

Using the ⊞ and the memory keys on your calculator will make it easier to calculate deductions and net pay.

LUND TOOL AND DIE PAYROLL REGISTER

For the Week Beginning April 19, 20— and Ending April 23, 20—   Paid April 29, 20—

NO.	NAME	MARITAL STATUS	EXEMP.	HOURS	REGULAR	OVERTIME	TOTAL	FIT	SS TAX	MEDICARE TAX	STATE TAX	UNION DUES	AAP	TOTAL	AMOUNT	CK. NO.
		EMPLOYEE DATA			EARNINGS					DEDUCTIONS					NET PAY	
111	Ahrend, M.	M	2	40	572 00	— 00						20 00				
112	Blase, G.	M	3	40	600 00	— 00						— 00				
117	Charlton, H.	S	0	42	544 00	40 80						20 00	— 00			
120	Dulong, R.	M	4	41	568 00	21 30						20 00				
121	Frish, U.	S	3	42	528 00	39 60						20 00				
124	Garth, M.	M	2	38	562 40	— 00						20 00	— 00			
130	Lennis, J.	M	4	44	496 00	74 40						20 00				
132	Mild, S.	S	0	40	576 00	— 00						— 00	— 00			
135	Milton, F.	M	3	41	548 00	20 55						20 00				
136	Pilcher, D.	S	2	37	558 70	— 00						20 00				
142	Popham, P.	S	1	40	564 00	— 00						20 00				
148	Popham, P.	M	5	43	512 00	57 60						20 00				
149	Ridley, J.	M	1	40	560 00	— 00						— 00				
152	Scott, W.	M	3	42	520 00	39 00						— 00				
154	Willard, J.	S	2	44	500 00	75 00						20 00				
	Totals															

328   Unit 6   Business and Consumer Math

Activities 6.4

Graph Activity

Jane Frederick's gross pay is $2,200 per month. She has $10 deducted for a dental plan and 4% of her gross pay for medical insurance. How much are her total health care deductions per month? _____

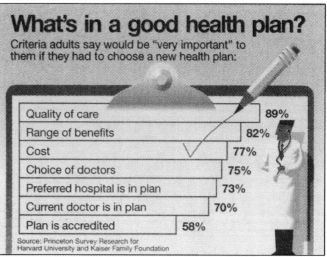

What's in a good health plan?
Criteria adults say would be "very important" to them if they had to choose a new health plan:

Quality of care	89%
Range of benefits	82%
Cost	77%
Choice of doctors	75%
Preferred hospital is in plan	73%
Current doctor is in plan	70%
Plan is accredited	58%

Source: Princeton Survey Research for Harvard University and Kaiser Family Foundation

By Anne R. Carey and Sam Ward, USA TODAY

Challenge Activity

Another method of computing Federal Income Tax (FIT) to be withheld is the percentage method, which can be utilized for any number of withholding allowances. Table 5, on page 330, shows the amount of *one* withholding allowance for various payroll periods. Tables 1 and 2, on page 330, are used to compute FIT for weekly and biweekly (every 2 weeks) pay periods. Social security and Medicare taxes are computed as before.

EXAMPLE

Hermione Gingold is paid biweekly. She is single, with 3 withholding allowances, and has earnings of $1,470 for the current pay period. Compute the amount of FIT to be withheld.

SOLUTION

Gross Earnings—		= $1,470.00
One allowance (from Table 5)—$103.85 × 3	=	− 311.55
Taxable Income—		= $1,158.45

From Table 2 (over $1,035 but not over $2,210), tax is $139.95
Plus 28% of excess over $1,035
$1,158.45 − $1,035 = $123.45
$123.45 × 0.28 = $34.57 + $139.95 = $174.52 FIT

Tables for Percentage Method of Withholding

TABLE 1—WEEKLY Payroll Period

(a) SINGLE person (including head of household)—

If the amount of wages (after subtracting withholding allowances) is:		The amount of income tax to withhold is:	
Not over $51		$0	
Over—	But not over—		of excess over—
$51	—$517 . .	15%	—$51
$517	—$1,105 . .	$69.90 plus 28%	—$517
$1,105	—$2,493 . .	$234.54 plus 31%	—$1,105
$2,493	—$5,385 . .	$664.82 plus 36%	—$2,493
$5,385	. . . . . .	$1,705.94 plus 39.6%	—$5,385

(b) MARRIED person—

If the amount of wages (after subtracting withholding allowances) is:		The amount of income tax to withhold is:	
Not over $124		$0	
Over—	But not over—		of excess over—
$124	—$899 . .	15%	—$124
$899	—$1,855 . .	$116.25 plus 28%	—$899
$1,855	—$3,084 . .	$383.93 plus 31%	—$1,855
$3,084	—$5,439 . .	$764.92 plus 36%	—$3,084
$5,439	. . . . . .	$1,612.72 plus 39.6%	—$5,439

TABLE 2—BIWEEKLY Payroll Period

(a) SINGLE person (including head of household)—

If the amount of wages (after subtracting withholding allowances) is:		The amount of income tax to withhold is:	
Not over $102		$0	
Over—	But not over—		of excess over—
$102	—$1,035 . .	15%	—$102
$1,035	—$2,210 . .	$139.95 plus 28%	—$1,035
$2,210	—$4,987 . .	$468.95 plus 31%	—$2,210
$4,987	—$10,769 . .	$1,329.82 plus 36%	—$4,987
$10,769	. . . . . .	$3,411.34 plus 39.6%	—$10,769

(b) MARRIED person—

If the amount of wages (after subtracting withholding allowances) is:		The amount of income tax to withhold is:	
Not over $248		$0	
Over—	But not over—		of excess over—
$248	—$1,798 . .	15%	—$248
$1,798	—$3,710 . .	$232.50 plus 28%	—$1,798
$3,710	—$6,167 . .	$767.86 plus 31%	—$3,710
$6,167	—$10,879 . .	$1,529.53 plus 36%	—$6,167
$10,879	. . . . . .	$3,225.85 plus 39.6%	—$10,879

Table 5. **Percentage Method—1998 Amount for One Withholding Allowance**

Payroll Period	One Withholding Allowance
Weekly	$ 51.92
Biweekly	103.85
Semimonthly	112.50
Monthly	225.00
Quarterly	675.00
Semiannually	1,350.00
Annually	2,700.00
Daily or miscellaneous (each day of the payroll period)	10.38

Using the tables above, or those in the Appendix, as appropriate, solve the following problem.

Reggie Peterson's weekly earnings are $1,088. He is married, and claims four withholding allowances. Compute his net, or take-home pay, including deductions for social security and Medicare taxes. _____

Internet Activity

The payroll deductions for medical plans usually include prescription drug plans. You pay a co-pay of $5–$10 and your company plan pays the rest. Some drugs cost as much as $50 or more for 1 month's supply. If you want information about the prescription drugs available or new drugs under development you can check this website.

http://pharminfo.com/drugpr/drugpr_mnr.ht

SKILLBUILDER 6.5

Taking Inventory

Learning Objectives

- **Complete simple inventory records.**
- **Verify the accuracy of inventory records with a physical inventory.**

Completing Inventory Records

A store's **inventory** is the quantity of goods or materials on hand. Taking inventory is the counting of goods or materials actually on hand. A **periodic inventory** requires counting the goods on hand at specific intervals, such as at year-end, semiannually, or quarterly. A **perpetual,** or running, **inventory** is one in which the counting is an ongoing process. In both cases, records, such as stock cards or computer runs, must be prepared and maintained. Computers and bar codes have made this task much easier, but not all businesses—especially smaller ones—lend themselves to a totally computerized operation. Knowing how much of a particular item is on hand is essential, and policies on when to reorder must be established in order to avoid being out of stock of an item.

> **MATH TIP**
>
> *An incorrect calculation on inventory items can result in the loss of a sale or overstocking of items.*

EXAMPLE

Baskets and More carries a large wicker basket that sells for $112.99.
Complete the following inventory card for the basket.

Date	Quantity Received	Quantity Sold	Balance
5/4			2
5/7	16	13	
5/12	10	15	
5/19	25	18	

SOLUTION

On each day, add the quantity received to the previous balance and
then subtract the quantity sold.

Date	Quantity Received	Quantity Sold	Balance
5/4			2
5/7	16	13	5
5/12	10	15	0
5/19	25	18	7

**Self-Check
Answer**
23

Self-Check
On May 31, Baskets and More received 35 of the wicker
baskets and sold 19. What was the inventory at the end of
the day?

Problems

Complete the four stock cards shown below. The minimum quantity is the reorder point. Circle any date on which any item falls below this amount.

1. Merchandise *Copy Paper, 5,000/case*

Minimum Quantity *36*

Merchandise *Copy Paper, 5,000/case*			
Minimum *36*			
Date	Quantity Received	Quantity Sold	Balance
3/2			30
3/4	15		
3/8	18	10	
3/15	25	36	
3/19	20	28	
3/26	12		
3/29	36	24	

2. Merchandise *$8\frac{1}{2}$ × 11 Ruled Pads, 12/pack*

Minimum Quantity *140*

Merchandise *$8\frac{1}{2}$ × 11 Ruled Pads, 12 pack*			
Minimum *140*			
Date	Quantity Received	Quantity Sold	Balance
3/3			155
3/9		110	
3/16	120		
3/18	120	90	
3/21		40	
3/25	180	50	
3/30		75	

3. Merchandise *#10 Envelopes, 50/pack*

Minimum Quantity *72*

Merchandise *#10 Envelopes, 50/pack*			
Minimum *72*			
Date	Quantity Received	Quantity Sold	Balance
3/1			81
3/5		36	
3/9	40		
3/13	114	72	
3/22		50	
3/27	50	40	

4. Merchandise *Staples, 5,000/box*

Minimum Quantity *70*

Merchandise *Staples, 5,000/box*			
Minimum *70*			
Date	Quantity Received	Quantity Sold	Balance
3/1			98
3/2	26		
3/11		48	
3/23	30	45	
3/24	59		
3/28		40	

5. Determine the total quantity in inventory for each item on the inventory sheet. Compare this to the minimum quantity indicated, and enter "Y" for "Yes" or "N" for "No" in the reorder column.

Item	Quan. at Loc. 1	Quan. at Loc. 2	Quan. at Warehouse	Total Quan.	Minimum Quan.	Reorder Y/N
Multifax #280388	8	2	7	_____	8	_____
Fax Modem #395731	4	6	5	_____	12	_____
Personal Copier #224402	1	5	3	_____	10	_____
Simpson Qt. Pr. #273052	2	1	0	_____	6	_____
Cordless Phone #284067	9	7	12	_____	30	_____
486SX/25 Comptr #322115	4	3	2	_____	10	_____

Physical Inventory

No matter what system is used for maintaining a running inventory, there are bound to be errors due to theft, breakage, mistakes in recording, and so on. To correct these errors, an actual count, or **physical inventory,** must be made of goods on hand. This is usually done at specified intervals but must be done whenever an accurate financial statement must be prepared.

EXAMPLE

Cramborski's Hardware lists 28 shovels in inventory that sell for $16.99. Is the total value of $473.72 listed on the inventory sheet correct?

SOLUTION

To find the total value, multiply the number in inventory by the unit costs:

$$\begin{array}{r} \$16.99 \\ \times\ 28 \\ \hline 135\ 92 \\ 339\ 8\ \ \\ \hline \$475.72 \end{array}$$

No, the total value on the inventory sheet should be $475.72.

Self-Check
Aaron's Art Store has 332 paintbrushes in stock that sell for $2.55. What value should be listed on the inventory sheet?

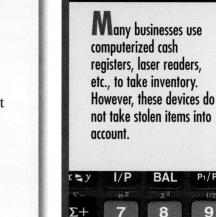

Calculator Tip

Many businesses use computerized cash registers, laser readers, etc., to take inventory. However, these devices do not take stolen items into account.

Self-Check Answer
$846.60

The two pages shown are part of the physical inventory taken at the end of the year. Check all extensions for accuracy, and enter any corrections in the space provided.

6.

PHYSICAL INVENTORY COUNT					CORRECTED TOTAL VALUE
SECTION Dinnerware			PART 12		
DATE Dec. 31, 20__			SHEET NO. 3		
COUNTED BY: F.H.	RECORDED BY: E.J.	COMPUTED BY: M.A.			
STOCK NO.	QUANTITY	DESCRIPTION	UNIT PRICE	VALUE	
24	20 sets	Lily	$ 27.88	$ 557.60	_____
33	23 sets	Stafford	62.00	1,326.20	_____
61	17 sets	Ironstone	89.00	1,513.00	_____
64	9 sets	Stoneware	54.40	504.27	_____
72	13 sets	Fuji	39.79	572.27	_____
80	39 sets	Pagoda	79.78	3,111.42	_____
93	11 sets	Bavaria	113.75	1,221.55	_____
		Total		$10,105.67	_____

7.

PHYSICAL INVENTORY COUNT

| SECTION | Floor coverings | | | | PART | 15 |
| DATE | Dec. 31, 20_ _ | | | | SHEET NO. | 5 |

COUNTED BY: M.R.   RECORDED BY: R.G.   COMPUTED BY: M.M.

STOCK NO.	QUANTITY	DESCRIPTION	AREA OF EACH	UNIT PRICE	VALUE OF EACH UNIT	TOTAL VALUE	CORRECTED TOTAL VALUE
103	12	Carstar vinyl	99 ft^2	$ 0.43/ft^2	$ 42.57	$ 510.84	_____
113	16	Bandon vinyl	99 ft^2	0.27/ft^2	26.73	427.68	_____
124	15	Econo vinyl	126 ft^2	2.17/yd.	30.38	4,231.50	_____
162	9	Runners	42 ft^2	0.38/ft^2	15.96	146.44	_____
171	20	Manor tile, 12 × 12 in.	45 ft^2	0.39/ft^2	17.55	351.00	_____
178	19	Guild tile, 12 × 12 in.	25 ft^2	1.25/ft^2	31.25	593.75	_____
202	7	Drew tile, 12 × 12 in.	25 ft^2	0.95/ft^2	23.75	166.25	_____
204	43	Nylon shag	12 yd^2	16.49/yd^2	197.88	8,508.84	_____
255	27	Nylon broadloom	180 ft^2	17.93/yd^2	358.60	9,682.20	_____
		Total				$24,618.50	_____

8. Compute the value of each of the following physical inventory for Sagnaro Bed and Bath.

Item	Description	Unit Price	Stock Quantity	Actual Count	Value
M-3564	Mini blind	8.99	123	132	_____
RB3672	Roll-up blind	17.00	57	57	_____
V7884	Vertical blind	69.99	23	21	_____
PV-3260	Pleated shade	11.95	12	38	_____
BBT592	Comforter, twin	59.99	5	5	_____
BBQ692	Comforter, queen	79.99	4	2	_____
FL813T	Sheet, twin flat	5.99	27	27	_____
FF813T	Sheet, twin fitted	5.99	31	26	_____
JTS81	Juvenile twin set	21.99	15	15	_____
				Total	_____

Business Applications 6.5

1. Complete the following inventory sheet for The Music Store order. Enter the quantity shipped and the balance remaining in inventory. If the quantity ordered is greater than the quantity in stock, ship what you have and back-order the difference.

Item	In Stock	Ordered	Shipped	Back Order	Balance
Port. Stereo	38	2	_____	_____	_____
13-in. Color TV	8	13	_____	_____	_____
VCR	12	2	_____	_____	_____
CD Changer	23	4	_____	_____	_____
Cassette Deck	5	8	_____	_____	_____
Port. CD Player	15	22	_____	_____	_____

2. Rachael's Tires compiled the following physical inventory sheet. Compute the value for each item based on the actual count.

Tire Item No.	Unit Price	Inventory Qty.	Actual	Value
P195/60HR14	$107.99	12	13	_____
P185/70SR13GT	53.99	21	21	_____
P215/75SR15	85.50	5	3	_____
33X12.50R15	115.95	2	2	_____
P165/80R13	33.75	35	39	_____
165SR13GT	42.00	7	8	_____
P195/70SR114BW	67.99	2	0	_____
LT235/85R16E	126.99	4	0	_____
P205/75R15XL	57.80	6	6	_____
			Total	_____

3. Complete the following inventory sheet for Mortar and Pestle, a manufacturer of pharmaceutical containers.

Item: Glass bottles			Location: Harrison Area 3.690e	
	Item	**Quantity**	**Unit Price**	**Total**
a.	60 ml	518,400 ea.	$2.16/gross	_____
b.	120 ml	4,464 gross	$518.40/gr.-gr.*	_____
c.	8 ounce	58,800 ea.	$1.50/doz.	_____
d.	475 ml	10,742 doz.	$0.155 ea.	_____

*gr-gr = gross-gross = 144 × 144

Activities 6.5

Graph Activity

If you wanted to keep an inventory of 100 copies of *Rainbow Six,* how many copies of *I Know This Much Is True* should you keep in hand? _____

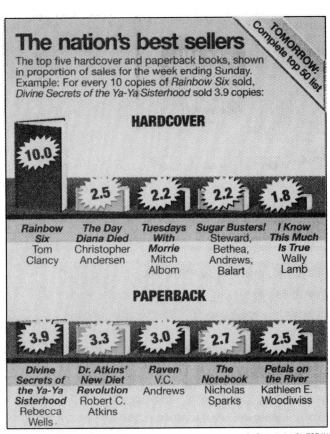

Challenge Activity

Make up a problem involving inventory. Include a table with columns titled Date, Quantity Received, Quantity Sold, and Balance. Be sure to include answers to your problem. Write your answers in a different color than the problem.

Internet Activity

Inventory is usually a distributor's largest asset and should contribute to the success of the business. What are three problems of inventory management?

http://www.effectiveinventory.com

Select "What Is Effective Inventory Management?"

SKILLBUILDER 6.6

Valuing Inventory

Learning Objectives

- **Compute inventory value based on cost price or selling price.**
- **Compute inventory value based on an average (weighted average) basis.**
- **Compute inventory value based on a first in, first out basis.**
- **Compute inventory value based on a last in, first out basis.**

Computing Inventory on Cost or Selling Price

When an inventory value is based on either cost price or selling price, the specific price of each lot of merchandise is used. The inventory value is found by multiplying the unit cost of each item by the number of items on hand. To find the total inventory value, add the inventory values of all items.

EXAMPLE

Compute the value of the following inventory based on the cost price.

SOLUTION

Quantity Purchased	Item	Unit Cost	On Hand	Inventory Value
27 cases	Tomatoes	$16.56/case	11 cases	$182.16
36 cases	Pork and beans	11.76/case	21 cases	246.96
21 cases	Asparagus	21.36/case	9 cases	192.24
6 cases	Tuna	30.00/case	4 cases	120.00
			Total	$741.36

Self-Check

The Big Barn carries three models of bicycles. They have 48 model A bikes in stock, selling for $118.75 each. They also have 27 of model B, selling for $199.99, and 11 of model C, selling for $244.99. Find the total inventory based on selling price.

Self-Check Answer
$13,794.62

Problems

Compute the inventory value of each item, and find the total inventory value for the following inventory sheets.

1. INVENTORY REPORT AS OF JULY 1, 20—

Dept. Jewelry

Store No. 83

Purchase Order No.	Quantity Purchased	Item	Unit Cost	On Hand	Inventory Value
316	48	Watch, mens quartz	$65.00	12	_____
321	72	Watch, digital	26.00	39	_____
358	24	Watch, ladies quartz	90.00	9	_____
402	36	Fashion watch	18.11	15	_____
473	75	Juvenile watch	12.00	4	_____
				Total	_____

2. INVENTORY REPORT AS OF JULY 1, 20—

Dept. Office Supplies

Store No. 83

Purchase Order No.	Quantity Purchased	Item	Unit Cost	On Hand	Inventory Value
516	8 doz.	Super Ball Pens	$21.46 doz.	3 doz.	_____
523	108 ea.	Halogen lamps	35.98 ea.	57	_____
524	10 doz.	Hi-Liters	16.20 doz.	5 doz.	_____
534	60 boxes	Computer paper	49.85 ea.	6	_____
547	84 boxes	DS/HD Disks/20	22.92 box	27	_____
				Total	_____

Computing Inventory Using the Average (Weighted Average) Method

When it is impractical to keep track of each purchase and its sale, the average, or weighted average, method is used. This method is used for a periodic inventory valuation and usually involves goods of relatively low value.

Name:_____ Date:_____

EXAMPLE

What is the value of the inventory of paintbrushes as of June 30?

Item	Paintbrush			Location		Max.	1200
				Bin No. 772		Min.	300

Date	Quantity In	Unit Cost	Req. No.	Quantity Out	Balance
20—					
5/15	350	$1.75			350
5/28			12	113	237
6/3	450	1.80			687
6/10	450	1.90			1,137
6/15			15	975	162
6/30	300	2.00			462

SOLUTION

First find the value of all units purchased at each price. Total the number of units and the inventory values.

$$\frac{No.\ of}{Units} \times \frac{Unit}{Cost} = \frac{Inventory}{Value}$$

$$350 \times \$1.75 = \$612.50$$
$$450 \times 1.80 = 810.00$$
$$450 \times 1.90 = 855.00$$
$$\underline{300 \times 2.00 = \quad 600.00}$$
$$1,550 \times \qquad = \$2,877.50$$

Then divide the total inventory value by the total number of units to find the average unit cost of each unit. Round to three decimal places.

$$\frac{Total\ Inventory}{Value} \div \frac{Total}{Units} = \frac{Average}{Unit\ Cost}$$

$$\$2,877.50 \div 1,550 = \$1.8564516 \rightarrow \$1.856$$

To find the inventory value on June 30, multiply the number of units on hand by the average unit cost.

$$\frac{Units}{on\ Hand} \times \frac{Average}{Unit\ Cost} = \frac{Inventory}{Value}$$

$$462 \times \$1.856 = \$857.472 \rightarrow \$857.47$$

Self-Check

What is the value of the inventory of measuring cups as of August 30?

Date	Quantity In	Unit Cost	Req. No.	Quantity Out	Balance
20—					
8/1	1,376	$0.88			1,376
8/10			2.46	810	566
8/15	1,500	$0.91			2,066
8/19			2.58	1,306	760
8/30	1,000	$0.93			1,760

Problems

Answer

3. _____

3. Compute the value of the following inventory as of July 30, using the weighted average method.

Item Fertilizer				
Date	Quantity In	Unit Cost	Quantity Out	Balance
20—				
3/30	360 bags	$ 7.20		360
4/5			315	45
4/13	300 bags	8.80		345
4/18			295	50
4/27	530 bags	8.00		580

Computing Inventory Using the First-In, First-Out (FIFO) Method

The FIFO method is also used for a periodic valuation and assumes that the first-, or earliest-, purchased merchandise is the first sold or used in filling orders. The most recently purchased goods remain in inventory.

Name: _____ Date: _____

EXAMPLE

Three different lots of merchandise are purchased at different prices: 10 units at $7 each, 5 units at $9 each, and 8 units at $8 each. The ending inventory is 15 units. What is the value of the ending inventory on a FIFO basis?

SOLUTION

Total purchases were 23 units (10 + 5 + 8 = 23). The ending inventory is 15 units; therefore, 8 units were sold (23 − 15 = 8). The assumption is that the 8 units sold during this time were all from the first lot purchased, so we work back from the most recently purchased lots to establish the inventory value.

Number of Units	Unit Cost		Inventory Values	
8	× $8.00	=	$64.00	
5	× 9.00	=	45.00	
2	× 7.00	=	14.00	(8 sold)
15			$123.00	

Self-Check

Compute the value of the ending inventory for the following using the FIFO method:

Beginning inventory	40 units at $20
First purchase	50 units at $21
Second purchase	50 units at $22
Third purchase	50 units at $23
Ending inventory	35 units

Student Success Hints

Studying for tests. Ask the instructor what types of problems will be on the test, how the problems will be graded, and how much time you will have to take the test.

Self-Check Answer
$805

Problems

4. Compute the value of the ending inventory for the following using the FIFO method.

Beginning inventory	42 units at $12
First purchase	35 units at $18
Second purchase	15 units at $13
Third purchase	28 units at $15
Ending inventory	48 units

Answer

4. _____

Computing Inventory Using the Last-In, First-Out (LIFO) Method

Another periodic method of valuing inventory is LIFO, which assumes that the last, or most recently purchased, merchandise is used to fill orders. The goods remaining in inventory would be from the lots purchased at the earliest dates.

Calculator Tip

When rounding to the nearest cent, set the decimal selector on the calculator on two places.

EXAMPLE

Three different lots of merchandise are purchased at the following prices: 15 units at $6 each, 10 units at $8 each, and 12 units at $7 each. The ending inventory is 16 units. What is the value of the ending inventory on a LIFO basis?

SOLUTION

Total purchases were 37 units (15 + 10 + 12 = 37). Ending inventory is 16 units; therefore, 21 units were sold (37 − 16 = 21). The assumption is that the goods sold are taken from the most recently purchased lots, so we work forward from the beginning inventory and the lots first purchased.

Number of Units		Unit Price		Inventory Values	
15	×	$6.00	=	$90.00	
1	×	8.00	=	8.00	(9 sold at $8)
					(12 sold at $7)
16				$98.00	

Self-Check Answer
$3,216

Self-Check

Compute the value of the ending inventory for the following using the LIFO method.

First purchase	60 units at $44
Second purchase	30 units at $48
Third purchase	25 units at $42
Fourth purchase	40 units at $40
Ending inventory	72 units

Answer

5. _____

Problems

5. Compute the value of the ending inventory for the following using the LIFO method.

Beginning inventory	168 units at $30
First purchase	372 units at $24
Second purchase	218 units at $16
Third purchase	127 units at $42
Ending inventory	545 units

Business Applications 6.6

Answers

1. _____

2. _____

3. _____

1. Cable Products' ending inventory of sound shields was 86 units. Beginning inventory was 35 units at $19.25 each. In May, 49 units were purchased at $18.75 each; in July, 60 units at $18.25 each; in October, 55 units at $19.75 each; in December, 62 units at $18.50 each. Compute the value of the ending inventory using the LIFO method.

2. Cable Products' ending inventory of safety lanterns was 212 units. Beginning inventory in January was 102 units at $24 each. In April 84 units were purchased at $23 each; in June, 78 units at $24.50 each; in August, 108 units at $22 each; in November, 80 units at $22.50 each. Compute the value of the ending inventory using the FIFO method.

3. Complete the balance for each of the transaction dates, and then compute the average inventory value as of May 30.

Item	Long-Handle Shovel			
Date	Quantity In	Unit Cost	Quantity Out	Balance
20—				
4/1	43	$6.19		43
4/7	154	6.19		_____
4/16			26	_____
4/23			41	_____
4/28	180	6.27		_____
5/5	40	6.25		_____
5/21			112	_____
5/30	100	7.11	56	_____

4. Reed Automotive had a beginning inventory on November 1 of five sets of tires that cost $230 each. On November 10, Reed purchased 10 more sets of tires for $220 each and on November 20 he purchased 10 more sets for $210 each. At the end of the month, seven sets of tires were left in ending inventory. What was the cost of ending inventory using the FIFO method?

5. Brighter Day Lamp Store had a beginning inventory on December 1 of seven floor lamps that cost $50 each. On December 5, the store manager purchased 10 more floor lamps for $52 each, and on December 22 another purchase was made of 15 floor lamps for $50 each. At the end of the month, eight floor lamps were left in ending inventory. What was the cost of the ending inventory using the LIFO method?

6. Office Systems, Inc. had a beginning inventory on January 1 of three desktop copiers that cost $300 each. On January 21, the manager purchased five more desktop copiers for $320 each. On January 30, the manager purchased another five desktop copiers for $330 each. At the end of the month two desktop copiers were left in ending inventory.

a. What was the cost of ending inventory using the FIFO method?

b. What was the cost of ending inventory using the LIFO method?

Activities 6.6

Graph Activity

PC Solutions, Inc. sells computers. Calculate the value of the ending inventory for the following using the FIFO method. _____

10 units at $1,200
10 units at $1,000
10 units at $ 900
ending inventory 12 units

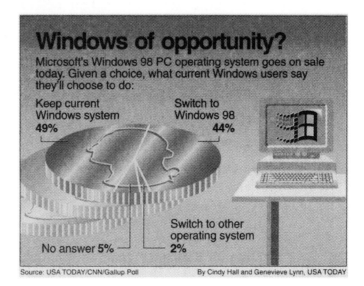

Windows of opportunity?

Microsoft's Windows 98 PC operating system goes on sale today. Given a choice, what current Windows users say they'll choose to do:

Keep current Windows system
49%

Switch to Windows 98
44%

Switch to other operating system
2%

No answer **5%**

Source: USA TODAY/CNN/Gallup Poll

By Cindy Hall and Genevieve Lynn, USA TODAY

Challenge Activity

Complete the following inventory card, and compute the ending inventory value using the Weighted Average, First-In, First-Out, and Last-in, First-Out methods.

Date	Quantity Received	Price	Quantity Sold	Balance Remaining
1/1		$12.80		3,100
1/29			2,200	_____
2/8	2,400	$14.72		_____
2/15	2,400	$15.46	1,800	_____
2/27			800	_____
3/15	3,600	$14.84	1,000	_____
3/30			1,500	_____

Internet Activity

Read more about effective inventory management and find out if you are making money.

http://www.effectiveinventory.com

Click on **articles** and select "Are You Making Money?" by Jon Schreibfeder.

Name:_____ Date:_____

SKILLBUILDER 6.7

Computing Depreciation by Straight-Line and Double Declining Balance Methods

Learning Objectives

- **Compute depreciation by the straight-line method.**
- **Compute depreciation by the double declining balance method.**

Depreciation

Depreciation is the loss incurred through the decline in value of property such as machinery, equipment, and buildings. It may also be due to obsolescence, which is the outdating of something because a new development or more modern item has replaced it. Four methods of computing depreciation are the straight-line method, the double declining balance method, the units-of-production methods, and the sum-of-the-years' digits method.

Straight-Line Method

The **straight-line** method for computing depreciation includes the following steps.

1. Determine the probable, or **estimated, life** of the property.

2. Determine the **trade-in,** or **scrap, value** of the item. This is an estimate of the item's worth at the end of its useful life.

3. Find the difference between the original cost and the trade-in value. This difference is the total depreciation. Divide the total depreciation by the estimated (probable) life to find the **annual depreciation.**

4. Divide the annual depreciation by the total depreciation to find the **annual rate of depreciation.**

EXAMPLE

A machine that cost $10,500 has a trade-in value of $1,575 at the end of 5 years. Find the annual depreciation and the rate of depreciation.

SOLUTION

$10,500   Original cost
−1,575   Trade-in value
 $8,925   Depreciation for the 5-year period
 $8,925   ÷ 5 = $1,785   Annual depreciation

Annual Depreciation	÷	*Total Depreciation*	=	*Rate of Depreciation*
$1,785	÷	$8,925	=	0.20, or 20%

Self-Check Answers
$2,262.50
0.10

Self-Check

A machine with an original value of $25,000 has a trade-in value of $2,375 after 10 years. Find (a) the annual depreciation and, (b) the rate of depreciation.

Problems

1. Complete the following.

Original Cost	–	Trade-in Value	=	Total Depreciation
a. $23,850	–	$2,425	=	_____
b. 19,959	–	4,500	=	_____
c. _____	–	16,910	=	112,520
d. 51,859	–	_____	=	47,537

2.

Total Depreciation	÷	Estimated Life	=	Annual Depreciation
a. $ 6,516	÷	3 y	=	_____
b. 87,645	÷	15 y	=	_____
c. _____	÷	9 y	=	4,927
d. 112,288	÷	___ y	=	7,018

3.

Annual Depreciation	÷	Total Depreciation	=	Rate of Depreciation
a. $ 1,528	÷	$ 6,112	=	_____
b. 18,235	÷	455,875	=	_____
c. 348	÷	_____	=	8.33%
d. _____	÷	22,500	=	15%

4. Complete the following straight-line depreciation schedule. Round the rate of depreciation to the nearest tenth of a percent where necessary.

Property	Original Cost	Trade-in Value	Estimated Life	Annual Depreciation	Rate of Depreciation
Truck	$ 49,600	$ 4,960	9 y	_____	_____
Building	1,490,000	None	_____	_____	3.3%
Equipment	198,400	38,884	_____	13,293	_____
Car	19,000	8,360	4 y	_____	_____
Computer	_____	5,500	5 y	7,900	20%

Double Declining Balance Method

The double declining balance method is used to compute the annual depreciation on the decreasing value of an item. The trade-in value is not deducted from the cost of the item. This method involves the following steps.

1. Divide 100% by the estimated life of the item to determine the straight-line annual rate of depreciation. For example, a 5-year useful life is equivalent to 20% a year (100% ÷ 5 = 20%).

2. Double this rate (2 × 20% = 40%).

3. Multiply the original cost of the item by the double rate to find the first year's depreciation.

4. Subtract the first year's depreciation from the original cost to determine the value remaining at that time, which is the **book value** for the first year.

5. Repeat steps 3 and 4 for each successive year, beginning with the book value at the end of the first year. Continue this procedure until the book value equals the trade-in value.

EXAMPLE

A computer that costs $3,200 is expected to last 8 years. Use the double declining balance method to find the book value at the end of 3 years.

SOLUTION

1. $100\% \div 8 = 12.5\%$

2. $12.5\% \times 2 = 25\%$, or 0.25

3.

1st Year Depreciation	=	Cost	×	Double Rate
$800	=	$3,200	×	0.25

4. $3,200   Cost
   − 800   First-year depreciation
   $2,400   Book value at end of first year

5. $2,400 × 0.25 = $600   Second-year depreciation
   $2,400 − $600 = $1,800   Book value at end of second year
   $1,800 × 0.25 = $450   Third-year depreciation
   $1,800 − $450 = $1,350   Book value at end of third year

Self-Check
A machine cost $12,000 and has an estimated life of 4 years. It is expected to have a trade-in value of $2,200 at that time. What is the book value of this machine at the end of the second year?

Student Success Hints

Studying for tests. Spend 25% of your time studying theory and formulas, and 75% of your time applying them.

Self-Check Answer
$3,000

5. A retail check-out terminal costing $12,500 has a trade-in value of $2,450 at the end of three years. What is the book value at the end of the second year?

Answers

5. _____

6. _____

6. A lift truck originally cost $181,500 and has an estimated trade-in value of $17,500 after 10 years. Find the book value of the truck at the end of the third year.

7. A piece of equipment cost $2,500 and is expected to last for 5 years. Using the double declining balance method, compute the depreciation for each of the 5 years, and show the book value at the end of the fifth year.

Year	Cost/Book Value	×	Rate of Depreciation	=	Depreciation for Year
	$2,500	×	_____	=	_____
	_____				
1. Book Value	_____	×	_____	=	_____
	− _____				
2. Book Value	_____	×	_____	=	_____
	− _____				
3. Book Value	_____	×	_____	=	_____
	− _____				
4. Book Value	_____	×	_____	=	_____
	− _____				
5. Book Value	_____				

8. If the equipment in Problem 5 had a trade-in value of $500, (a) what would the amount of depreciation be at the end of year 4, and (b) what would the book value be at that time?

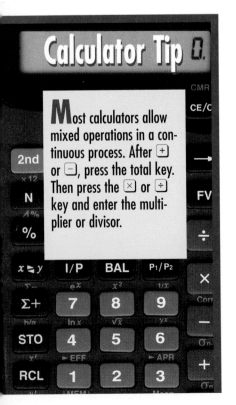

9. A computer that originally cost $13,500 is estimated to have a trade-in value of $2,500 at the end of 6 years. Determine its book value at the end of the fourth year.

Answers

8. a. _____

b. _____

9. _____

© Glencoe/McGraw-Hill

Name:_____ Date:_____

Business Applications 6.7

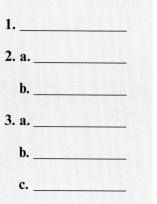

Answers

1. _____

2. a. _____

   b. _____

3. a. _____

   b. _____

   c. _____

1. The annual rate of depreciation on a warehouse was estimated to be 3.2% of the original cost. If the warehouse cost $2,850,000 what was the annual depreciation?

2. A heavy-duty copy machine cost Loeb Sportswear $108,000 installed and is estimated to have a 15-y useful life, with a salvage value of $6,000. What are (a) the annual depreciation and (b) the rate of depreciation?

3. Pat's Floral Shop bought a mini-van costing $27,200. After 4 years, the van had a resale value of 44% of the original cost.

   a. Find the resale value.

   b. Find the annual depreciation.

   c. Find the rate of depreciation to the nearest tenth of a percent.

4. Find the depreciation for the first year on an asset that cost $88,000, has a useful life of 4 years and no salvage value. Use the straight-line method.

5. Find the depreciation for the first year on an asset that cost $22,000, has a useful life of 5 years and a salvage value of $2,000. Use the straight-line method.

6. Find the depreciation for the second year on an asset that cost $17,500, has a useful life of 4 years and a salvage value of $1,500. Use the double-declining balance method.

7. Find the depreciation for the second year on an asset that cost $44,000, has a useful life of 8 years and a salvage value of $4,000. Use the double-declining balance method.

Activities 6.7

Graph Activity

R & E Properties, Inc. purchased an apartment complex for $480,000. The complex has a 40-year life and no trade-in value for depreciation purposes. Calculate the annual depreciation using the straight-line method. _____

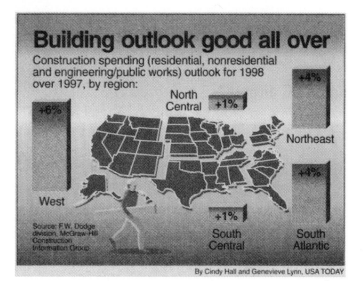

Building outlook good all over

Construction spending (residential, nonresidential and engineering/public works) outlook for 1998 over 1997, by region:

+6% West

North Central +1%

+4% Northeast

+4% South Atlantic

+1% South Central

Source: F.W. Dodge division, McGraw-Hill Construction Information Group

By Cindy Hall and Genevieve Lynn, USA TODAY

Challenge Activity

Dubrow Freres purchased a production machine for $84,000 with a projected useful life of 16 years, at which time it will have a scrap value of $2,000.

a. Using the straight-line method of depreciation, what will the book value be in year 10? _____

b. Compare the book value in year 5 using the straight-line and double declining balance methods. _____

_____

Internet Activity

As a small business owner you have a small budget for equipment. You must buy a copy machine to save on copying costs. Go to

http://www.discountcopier.com

and check the prices for used copiers. Find out the depreciation for your favorite copier.

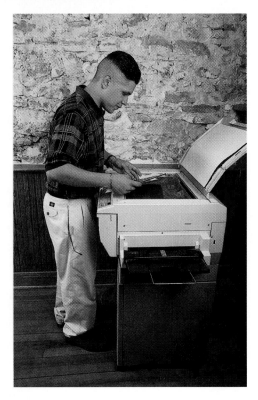

SKILLBUILDER 6.8

Computing Depreciation by Units of Production and Sum-of-the-Years' Digits Methods

Learning Objectives

- **Compute depreciation using units of production, either units produced or hours in operation.**
- **Compute depreciation using the sum-of-the-years' digits method.**

Units of Production Method

The useful life of equipment is sometimes estimated in terms of the total amount of work performed either in the number of units produced or the number of hours of operation. This way of computing depreciation is known as the **units of production** method.

If a machine wears out after it has produced 50,000 units, its **production life** is 50,000 units.

The depreciation for the year based on units produced is determined by the following steps:

1. Determine total depreciation by subtracting the trade-in value from the cost.

2. Determine the unit depreciation by dividing total depreciation by the production life in units.

3. Find the depreciation for the year by multiplying the units produced by the unit depreciation rate.

Depreciation can also be based on hours of operation. To find depreciation based on hours of operation, follow these steps:

1. Find the total depreciation.

2. Divide total depreciation by production life in hours to find hourly depreciation.

3. Find the depreciation for the year by multiplying the hours used by the hourly depreciation.

EXAMPLE

a. If a machine costs $28,600, has a production life of 50,000 units, and has an estimated trade-in value of $1,400, determine the depreciation in a year in which it produces 4,750 units.

b. If the machine described in part (a) has a useful production life of 17,500 hours and was operated for 825 hours during one year, find the depreciation based on hours of operation.

SOLUTION

a. 1.

$28,600	Cost
− 1,400	Trade-in value
$27,200	Total depreciation

2.

$$\textit{Unit Depreciation} = \textit{Total Depreciation} \div \textit{Production Life in Units}$$
$$\$0.544 = \$27,200 \div 50,000$$

3.

$$\textit{Depreciation for Year} = \textit{Units Produced} \times \textit{Unit Depreciation Rate}$$
$$\$2,584 = 4,750 \times \$0.544$$

b. 1. Total depreciation = $27,200 [part (a)].

2.

$$\textit{Hourly Depreciation} = \textit{Total Depreciation} \div \textit{Production Life in Hours}$$
$$\$1.554285 \rightarrow \$1.55429 = \$27,200 \div 17,500$$

3.

$$\textit{Depreciation for Year} = \textit{Hours Used} \times \textit{Hourly Depreciation}$$
$$\$1,282.289 \rightarrow \$1,282.29 = 825 \times \$1.55429$$

Self-Check Answers
a. $1,776
b. $2,571.43

Self-Check

The production life of a machine costing $78,000 is estimated at 180,000 units. Its trade-in value is estimated at $6,000, and its estimated production life in hours is estimated at 42,000 h.

a. If yearly production is 4,440 units, find the depreciation.

b. If it is used 1,500 h in one year, find the depreciation.

Problems

1. The production life of a machine is estimated at 180,000 units. The machine costs $78,000 and has an estimated trade-in value of $6,000. Complete the depreciation schedule below.

Year	Production	Depreciation
1	5,088 units	_____
2	3,510 units	_____

2. Compute the depreciation for the same machine if its estimated production life is 42,000 hours. It was used for the periods listed below. Round the hourly depreciation to 5 decimal places.

Year	Hours of Use	Depreciation
1	600	_____
2	2,200	_____

3. A machine that cost $17,700 has an estimated production life of 25,000 hours, with no trade-in value at the end of that time. If the machine was used 680 hours last year, find the depreciation for that year. Carry out the hourly depreciation to 3 decimal places.

4. An automatic casting machine has a production life of 3,000,000 units. The machine cost $75,000 and has an estimated trade-in value of $1,500. Find the depreciation if 480,000 units were produced the first year. Carry out the unit depreciation rate to 4 decimal places.

5. What is the book value at the end of the first year for the casting machine in Problem 4?

Answers

3. _____

4. _____

5. _____

The Sum-of-the-Years' Digits Method

The **sum-of-the-years' digits** method assumes the greatest depreciation for the years when an article is newest and its value is likely to drop the most. The rate of depreciation decreases each year. To compute depreciation using this method, follow these steps.

1. Estimate the useful life of the article in years.
2. Find the sum of all the digits of the years of estimated useful life. For example, for a machine with an estimated useful life of 4 years, the sum of the digits would be $1 + 2 + 3 + 4 = 10$.
3. Form a depreciation fraction by using the sum of the digits (in this case, 10) as the denominator of the fraction. The numerator is the digit of each of the individual years in reverse order: year 1 = $\frac{4}{10}$, year 2 = $\frac{3}{10}$, year 3 = $\frac{2}{10}$, and year 4 = $\frac{1}{10}$.
4. Subtract the estimated trade-in value from the original cost to find the total depreciation.
5. Multiply the total depreciation by the depreciation fraction to determine the depreciation for the first year.
6. Subtract the depreciation from the original cost to obtain the book value at the end of the first year.
7. Repeat steps 5 and 6, as necessary, to find the depreciation and book value for the required years.

MATH TIP

The sum of n years can be found by using the formula $n(n + 1)/2$. For 5 y, the sum is $5(5 + 1)/2 = 30/2 = 15$.

EXAMPLE

A truck costs $34,800. After 5 years, it will have a trade-in value of $5,900. Construct a depreciation schedule using the sum-of-the-years' digits method.

SOLUTION

$1 + 2 + 3 + 4 + 5 = 15$

$34,800	Cost
− 5,900	Trade-in value
$28,900	Total depreciation

Year	Depreciation	Book Value at End of Year
1	5/15 × $28,900 = $9,633.33	$34,800.00 − $9,633.33 = $25,166.67
2	4/15 × 28,900 = 7,706.67	25,166.67 − 7,706.67 = 17,460.00
3	3/15 × 28,900 = 5,780.00	17,460.00 − 5,780.00 = 11,680.00
4	2/15 × 28,900 = 3,853.33	11,680.00 − 3,853.33 = 7,826.67
5	1/15 × 28,900 = 1,926.67	7,826.67 − 1,926.67 = 5,900.00*
*Trade-in value		

Name:_____ Date:_____

Self-Check
A delivery truck was purchased at a cost of $24,000. The salvage value after 4 y was expected to be $3,000. Compute the amount of depreciation for each year.

Self-Check Answers
Year 1 $8,400
Year 2 $6,300
Year 3 $4,200
Year 4 $2,100

Problems

6. If the useful life of an item is 8 y, find

   a. The sum-of-the-years' digit

   b. The fraction for the first year's depreciation.

Answers

6. a. _____

   b. _____

7. a. _____

   b. _____

   c. _____

7. A solar heating installation cost $30,500 and has an estimated life of 20 y, with a trade-in value of $2,500 at the end of that time. Find the depreciation for (a) the first year, (b) the second year, and (c) the third year.

8. Complete the following depreciation table for an industrial grinder that cost $4,200 and will have a salvage value of $750 at the end of 4 y.

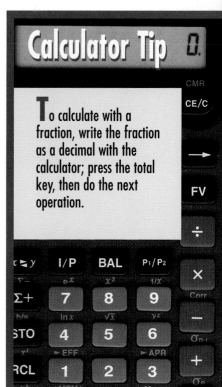

Calculator Tip 0.

To calculate with a fraction, write the fraction as a decimal with the calculator; press the total key, then do the next operation.

Year	Total Depreciation	×	Depreciation Rate	=	Annual Depreciation	Book Value
					0	$4,200
1	_____	×	_____	=	_____	_____
2	_____	×	_____	=	_____	_____
3	_____	×	_____	=	_____	_____
4	_____	×	_____	=	_____	_____

Business Applications 6.8

1. Coronet Mfg. purchased a machine that cost $86,800 and had a trade-in value of $6,500 after producing 550,000 units. Complete the following depreciation schedule.

Year	Units Produced	×	Depreciation per Unit	=	Depreciation for the Year	Cumulative Depreciation	Book Value
					0	0	$86,800
1	40,000	×	_____	=	_____	_____	_____
2	160,000	×	_____	=	_____	_____	_____
3	60,000	×	_____	=	_____	_____	_____
4	36,000	×	_____	=	_____	_____	_____
5	32,000	×	_____	=	_____	_____	_____

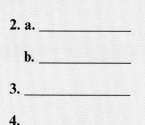

Answers

2. a. _____

    b. _____

3. _____

4. _____

2. The State Department of Transportation bought a road grader that cost $165,000 and is expected to be operated for 50,000 h before it must be refurbished. At that time, its engine will be rebuilt at a cost of $42,000. What is (a) the cost per hour of rebuilding as compared to (b) the cost per hour to replace the machine if its replacement cost is still $165,000?

3. A courier vehicle owned by Roadrunner Express was driven 35,000 mi in its first year of operation. The depreciation charge for the year was $4,200. The expected useful life of the vehicle is 150,000 mi, with a trade-in value of $500. What was the original cost of the vehicle?

4. A $20,000 car is bought and depreciated over 5 years. Compute the book value after 2 years using the sum-of-the-digits method.

Activities 6.8

Graph Activity

Ryan's Taxi Service purchased a new car to use as a taxi. The cost of the car was $22,000 and the trade-in value $2,000. Ryan's Taxi Service estimates the car will run 160,000 miles and uses the units-of-production method of depreciation. If the car is driven 42,000 miles the first year, find the annual depreciation. _____

Car of the future challenge

Winners are named today in the 1998 FutureCar Challenge, a government/industry contest[1] for universities to develop a family car that's fuel-efficient (80 mpg this year). Annual savings if cars averaged the 1996 winner's 42 mpg:

Source: FutureCar Challenge by the Department of Energy and the U.S. Council for Automotive Research

$9 billion savings on oil imports

$57.5 billion savings on gas

600 million barrels of imported oil not needed

1 – Schools get grant and mid-size car to re-engineer

By Cindy Hall and Quin Tian, USA TODAY

Challenge Activity

A new computer system was installed at a cost of $90,000. The system was projected to have a 5 year useful life before it would be replaced, at which point it could be sold for $11,500. At the end of the third year new technology made the system obsolete, so that its resale value was reduced to $20,000. If the system was replaced at that point, how much would the company lose? _____

Internet Activity

Do you need to brush up on your graphing skills? This internet site has a series of short interactive tutorials that introduce the basic concepts of graph theory.

http://www.utm.edu/departments/math/grap

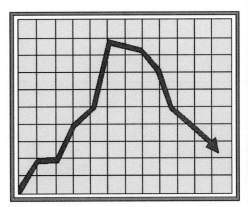

UNIT 7
INTEREST AND DISCOUNTS

In this unit, you will study the following Skillbuilders:

In this unit we discuss how to compute simple and compound interest using a variety of strategies, as well as how to discount noninterest-bearing and interest-bearing notes. Finally, we learn how to compute cash and trade discounts.

According to the figure shown here, 41% percent of Americans responding to a survey believe that rising interest rates affect them negatively. What are some reasons for this? How are you affected by rising interest rates?

If Interest Rates Rise

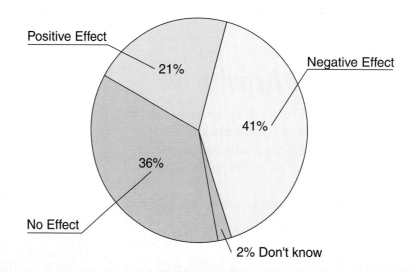

HOUSE
SOLD
SALE
782-0988

MATH CONNECTIONS

Bakery Owner

A former baker, Kelly, decides to open his own bakery. As an entrepreneur, Kelly must be able to market products, bake bread and pastries, serve customers, and take care of the business' finances. Kelly will need to project earnings and cash required to start the business and supplement earnings until a profit is earned.

Math Application

Kelly decides that he would like to expand his business, but that delivery of the bakery items would be necessary. Kelly decides to price some vans and trucks. Kelly finds a used van priced at $11,000. He can finance the van for 6% interest over 3 years. If Kelly makes a 15% down payment, what will be his monthly payments?

First, compute the loan amount.

$$\text{Price times 15\%} = \text{down payment}$$

Now compute the loan amount.

$$\text{Purchase price} - \text{down payment} = \text{loan amount}$$

Compute the interest and monthly payment.

$$\text{1st month interest} = \text{Principal} \times \text{Rate} \times \text{Time}$$

Interest on unpaid balances.

$$\frac{\text{1st month interest} \times (\text{Number of payments} + 1)}{2}$$

$$\text{Monthly Payment} = (\text{Principle} + \text{Interest})/\text{Number of months toward loan.}$$

Critical Thinking Problem

Kelly can lease a new van for $250 a month for 36 months, or purchase a last year's van for $300 a month for 60 months. What are some issues that Kelly should consider before making a final decision?

SKILLBUILDER 7.1

Computing Simple Interest by Formula

Learning Objectives

- **Compute interest using the interest formula and either a 12-month year or a banker's year.**
- **Find the exact time between two dates.**
- **Compute interest using the ordinary interest method.**

Simple Interest and the Banker's Year

Interest can be considered the rent paid to borrow money. The **principal** is the amount of money borrowed, and the **rate** is the percent of interest charged.

Time is the period of time for which the money is on loan. Since many loans are for less than one year, time may be expressed as a fraction of the year in terms of days, weeks, or months. Use this formula to compute interest.

Interest = Principal × Rate × Time
or

$$I \quad = \quad P \quad \times \quad R \quad \times \quad T$$

When the time of the loan is expressed in days, the number of days of the loan is divided by the number of days in a year. For purposes of computing interest, a **banker's year** or **commercial year** of 360 is used.

Find the interest on $1,269 at 9% for 5 mo.

SOLUTION

$$I = P \times R \times T$$
$$\$47.59 = \$1,269 \times 0.09 \times \frac{5}{12}$$

Find the interest on $3,200 at 11% for 90 days, using the banker's year.

SOLUTION

$$I = P \times R \times T$$
$$\$88 = \$3,200 \times 0.11 \times \frac{90}{360}$$

Self-Check Answers
$320
$195.50

Self-Check

a. Find the interest on $8,000 at 6% for 8 mo.

b. Find the interest on $4,600 at 8.5% for 180 d, using the banker's year.

Problems

Compute the interest in these problems using the formula $I = P \times R \times T$. Use the banker's year (360 d) when days are indicated. Round percents to the nearest tenth of a percent, where necessary.

1. $1,850 at 9% for 90 d

2. $600 at 11% for 75 d

3. $3,500 at 6% for 145 d

4. $7,500 at $5\frac{1}{2}$% for 5 mo

5. $11,000 at $7\frac{3}{4}$% for 212 d

6. $3,853 at $8\frac{1}{8}$% for 1 y 4 mo

7. $5,780 at $4\frac{7}{8}$% for 182 d

8. $25,167 at 8% for 47 d

9. $1,927 at $6\frac{3}{8}$% for 135 d

10. $970 at $5\frac{1}{8}$% for 30 d

11. $11,680 at $9\frac{1}{4}$% for 11 mo

12. $7,827 at $7\frac{5}{8}$% for 7 mo

Answers

1. _____
2. _____
3. _____
4. _____
5. _____
6. _____
7. _____
8. _____
9. _____
10. _____
11. _____
12. _____

Finding the Exact Time

The exact time is found by counting the actual number of days between any two dates, excluding the first date but including the last date.

When counting days, remember that the month of February has 29 d instead of 28 during a leap year. A leap year occurs every 4 yr. If the number of a year is evenly divisible by 4, it is a leap year. (There is an exception. If the year ends in at least two zeros, for example, 1800, 1900, or 2000, it must be divisible by 400 to be a leap year.) For example,

$$2000 \div 400 = 5 \quad 2000 \text{ is a leap year}$$

EXAMPLE

Find the exact number of days between December 12, 1999, and March 28, 2000.

SOLUTION

From December 12   19 d left in December
                                    31 d in January
                                    29 d in February (leap year)
To March 28   28 d in March
                                107 d, exact time

Self-Check
Answer
57

Self-Check
Find the exact number of days between January 5, 1999 and March 3, 1999

Problems

Find the exact number of days between the following dates.

13. May 5 to September 17

14. October 23 to January 8

15. July 7 to November 29

16. January 4 to March 24

17. June 15 to August 23

18. September 1 to December 29

Compute the interest due on the following loans using the banker's year and exact time.

19. $2,400 at $7\frac{1}{2}$% from January 12 to May 27

Answers

13. _____

14. _____

15. _____

16. _____

17. _____

18. _____

19. _____

20. $600 at 8% from August 14 to December 1

21. $11,750 at $9\frac{3}{4}$% from April 3 to May 27

22. $1,241 at $4\frac{7}{8}$% from December 21 to March 18

23. $32,975 at $7\frac{7}{8}$% from April 5 to July 23

24. $1,490 at 6% from January 4 to February 15

25. $9,875 at $10\frac{1}{8}$% from June 21 to November 5

26. $28,653 at $9\frac{5}{8}$% from February 21 to May 5

27. $2,050 at $8\frac{3}{4}$% from October 15 to January 5

28. $17,962 at $9\frac{1}{8}$% from November 3 to March 3

29. $8,125 at $5\frac{3}{4}$% from February 12 to August 31

30. $4,900 at $6\frac{7}{8}$% from July 9 to September 19

Ordinary Interest

Ordinary interest is frequently used in business because it is easier to count the time and determine the due date of loans. Ordinary interest assumes that a year consists of 12 mo of 30 d each. To find the date on which a loan is due, count forward the number of months of the loan. The loan will be due on the same day of the month as the loan began.

If the month the loan is due does not contain the same last day number as the month when the loan was made, then the last day of the month is used.

EXAMPLE

Find the due date of a 3-mo loan made on April 15.

SOLUTION

April 15 to May 15	1 month
May 15 to June 15	1 month
June 15 to July 15	1 month
	3 months

The loan is due on July 15.

EXAMPLE

Find the due date of a 2-mo loan made on July 31.

SOLUTION

July 31 to August 31   1 month
August 31 to September 30   1 month
                              2 months

The loan is due on September 30.

The method used to count time may cause a difference of a day or two in the length of the loan, which will affect the amount of interest earned. For example, the 2-mo loan in this example was made on July 31 and was due on September 30. By ordinary interest, the time would be counted as follows: 2 mo × 30 d = 60 d. By the exact time method, the exact count of days from July 31 to September 30 would be 61 d (31 d in August plus 30 d in September).

Self-Check

Find the due date of a 4-mo loan made on May 31.

Self-Check Answer
September 30

Problems

Determine the date on which each of the following loans will be due.

31. A 2-mo loan made on July 31

32. A 3-mo loan made on November 30

33. A 7-mo loan made on March 15

34. An 11-mo loan made on April 21

Using the ordinary interest method, determine (a) when the following loans are due and (b) the amount of interest to be paid.

35. $1,650 at 8% for 2 mo from April 15.

36. $4,269 at $5\frac{1}{2}$% for 6 mo from August 31, 1999.

37. $12,775 at $9\frac{1}{2}$% for 3 months from June 4

38. $21,520 at $7\frac{1}{8}$% for 4 months from July 31

Answers

31. _____
32. _____
33. _____
34. _____
35. a. _____
    b. _____
36. a. _____
    b. _____
37. a. _____
    b. _____
38. a. _____
    b. _____

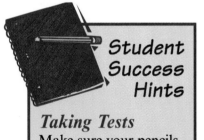

Student Success Hints

Taking Tests
Make sure your pencils are sharpened, have a good eraser handy, and if calculators are allowed, check the batteries.

39. $38,995 at $6\frac{3}{4}$% for 11 months from November 30

40. $6,150 at $8\frac{5}{8}$% for 5 months from January 18

41. $15,385 at $7\frac{1}{4}$% for 9 months from August 23

42. $9,775 at $6\frac{7}{8}$% for 8 months from March 17

Answers

39. a. _____

   b. _____

40. a. _____

   b. _____

41. a. _____

   b. _____

42. a. _____

   b. _____

Calculator Tip 0.

Set the decimal-point selector at two when computing the interest amount.

CMR

CE/C

→

FV

%   CST   SEL   MAR   ÷

I/P   BAL   P₁/P₂   ×

Σ+   7   8   9   Corr

STO   4   5   6   −

RCL   1   2   3   +

Business Applications 7.1

Answers

1. a. _____

   b. _____

2. _____

3. a. _____

   b. _____

4. a. _____

   b. _____

1. McGill Consultants borrowed $12,500 to purchase new office furniture. The bank quoted an interest rate of $9\frac{1}{2}\%$ for 180 d. Using the banker's year, compute (a) the amount of interest McGill will be charged and (b) the total amount to be repaid.

2. On January 30 Knutsen's Insurance borrowed $2,800 for 1 mo at a rate of 9%. (a) Is the company paying more or less for this loan on an ordinary interest basis, as compared to paying interest for the exact number of days of the loan? (b) How much more or less will it pay?

3. Bernadette's Audiology applied for a commercial loan in the amount of $42,800 for a period of 2 y. Because it was an unsecured loan, she was quoted a rate of $10\frac{1}{2}\%$. If the loan was secured by pledging company assets, the rate would be reduced by 2 percentage points. How much would each loan rate cost her?

4. On August 15, Monica obtained a 9 month loan of $7,850 at $6\frac{5}{8}\%$.
(a) When is the loan to be repaid? (b) How much will Monica repay?

Activities 7.1

Graph Activity

If a $10,000 certificate of deposit paid the same rate of interest as the Treasury bond 30-year yield, how much interest would this investment earn in one year? _____

Monday markets

Index	Close		Change
Nasdaq composite	1839.21	⬇	7.56
Standard & Poor's 500	1083.14	⬇	6.31
Treasury bond, 30-year yield	5.62%	—	unch.

Market Scoreboard, 3B; currencies, 4B

9:30 a.m. **8598**

Dow Jones industrial average
⬇ 23.17   4:00 p.m. **8575**

Challenge Activity

Suppose you want to borrow $7,550 to buy a used car. You want to pay it off in 15 months. Call a bank to find the current lowest interest rate.

a. How much interest will you pay?

b. What is the amount of your payment?

c. Can you afford the payment?

d. What amount of time makes the loan payment more affordable?

Internet Activity

What happens if you want to pay off a simple interest loan early? Go to the Infoseek search engine and type in the key words Simple Interest Loan.

SKILLBUILDER 7.2

Using a Simple Interest Table

Learning Objectives

- **Compute simple interest using a formula.**
- **Compute simple interest using a table.**
- **Find the principal, rate, or time as the unknown.**

Finding Simple Interest

Simple interest can be computed by using a table such as the Simple Interest Table shown in the Appendix. This table shows the interest earned for $100 at various rates of interest and for different periods of time. To use the simple interest table, follow these steps.

1. Find the appropriate Rate-of-Interest column at the top of the table.

2. Find the appropriate number of days or months of interest in the Time column.

3. Locate the point at which the Rate-of-Interest column and the Time column meet. This point indicates the amount of interest earned by $100 at that rate for that period of time.

4. Divide the principal by 100 (move the decimal point two places to the left) to determine the number of $100 units contained in the principal.

5. Multiply the interest on $100 by the factor obtained in step 4 to determine the amount of interest earned.

   To determine the interest factor for time periods or interest rates not shown on the table, factors may be combined. The simplest combinations should always be used when combining factors.

EXAMPLE

Find the interest earned on $1,729.85 for 17 d at a rate of $6\frac{1}{2}\%$. Round the answer to the nearest cent.

SOLUTION

Select the $6\frac{1}{2}\%$ column, and then move down to row 17. The factor of 0.3069 is the interest earned on $100 for 17 d at $6\frac{1}{2}$ percent.

$$\begin{array}{ccc} & & \textit{Interest} \\ \textit{Interest} & = & \textit{Loan} \times \textit{Factor} \\ \$5.3089096 \rightarrow \$5.31 & = & \$17.2985 \times 0.3069 \end{array}$$

Computing the same interest by the interest formula method gives the following result.

$$\begin{array}{ccccc} I & = & P & \times R & \times & T \\ \$5.309678 \rightarrow \$5.31 & = & \$1,729.85 & \times 0.065 & \times & \dfrac{17}{360} \end{array}$$

This demonstrates that the table gives a sufficiently accurate result.

EXAMPLE

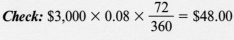

Using the simple interest table, find the interest earned on $3,000 at 8% for 72 days.

SOLUTION

$$8\% \quad = 2\tfrac{1}{2}\% \qquad\qquad + \quad 5\tfrac{1}{2}\%$$

72 days   = 2 months (60 days) + 12 days

$$\begin{array}{llll} 2 \text{ months} = & 60 \text{ days at } 2\tfrac{1}{2}\% & = 0.4167 \\ & 60 \text{ days at } 5\tfrac{1}{2}\% & = 0.9167 \\ & 12 \text{ days at } 2\tfrac{1}{2}\% & = 0.0833 \\ & \underline{12 \text{ days at } 5\tfrac{1}{2}\% \;\; = 0.1833} \\ & 72 \text{ days at } 8\% & = 1.6000 \end{array}$$

Interest = Loan divided by 100 × Interest Factor
$$\$48.00 \; = \$30.00 \qquad\qquad \times \qquad 1.6000$$

Check: $\$3,000 \times 0.08 \times \dfrac{72}{360} = \$48.00$

**Self-Check
Answer**
$34.72

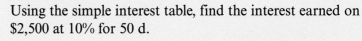

Self-Check

Using the simple interest table, find the interest earned on $2,500 at 10% for 50 d.

Problems

Use the Simple Interest Table shown in the Appendix to compute the interest charges on the following loans. You will have to determine the exact number of days for Problems 13 through 16.

1. $2,500 at $7\frac{1}{2}$% for 48 d

2. $1,725 at 7% for 180 d

3. $8,000 at $8\frac{1}{2}$% for 86 d

4. $426 at $5\frac{1}{2}$% for 75 d

5. $800 at 8% for 112 d

6. $2,980 at $9\frac{1}{2}$% for 99 d

7. $15,265 at $8\frac{1}{2}$% for 205 d

Answers

1. _____
2. _____
3. _____
4. _____
5. _____
6. _____
7. _____

8. $7,458 at 12% for 141 d

9. $1,850 at 10% for 54 d

10. $4,200 at 9% for 32 d

11. $11,875 at $7\frac{1}{2}$% for 93 d

12. $1,183 at $9\frac{1}{2}$% for 67 d

13. $3,850 at 8% from October 19 to April 17

14. $504 at $7\frac{1}{2}$% from November 20 to March 2

15. $8,250 at 5% from Feb. 11, 2000 to Nov. 24, 2000

16. $402 at 9% from May 20 to September 9

Finding the Principal, Rate, or Time

The following chart can be used to find the formula for the principal, rate, or time.

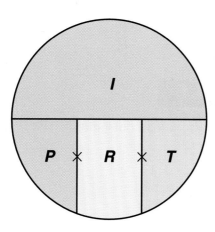

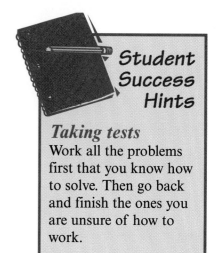

Student Success Hints

Taking tests
Work all the problems first that you know how to solve. Then go back and finish the ones you are unsure of how to work.

To find the **principal,** cover the letter *P.* The remaining letters indicate that the principal is found by dividing the interest by the product of the rate multiplied by the time.

$$P = \frac{I}{R \times T}$$

To find the **rate,** cover the letter *R* on the chart. The remaining letters indicate that the rate is found by dividing the interest by the product of the principal multiplied by the time.

$$R = \frac{I}{P \times T}$$

To find the **time,** cover the letter *T* on the chart. The remaining letters indicate that the time is found by dividing the interest by the product of the principal multiplied by the rate.

$$T = \frac{I}{P \times R}$$

The resulting number represents a fraction of a year and must be multiplied by 360 or 365 to determine the number of days for which the interest was paid.

EXAMPLE

If $42 interest was earned at a rate of 8% for 54 d, what was the amount of the loan?

SOLUTION

The amount of the loan, or the principal, is the unknown. Change 8% to a decimal fraction, and substitute the known quantities in the formula.

$$P = \frac{I}{R \times T}$$

$$= \frac{\$42}{0.08 \times \frac{54}{360}} = \frac{\$42}{0.012}$$

$$= \$3,500$$

EXAMPLE

For how many days was a loan outstanding if $1,750 earned $31.50 interest at a rate of 9%?

SOLUTION

$$T = \frac{I}{P \times R}$$

$$= \frac{\$31.50}{\$1,750 \times 0.09} = \frac{\$31.50}{\$157.50}$$
$$= 0.2 \text{ year or } 0.2 \times 360 \text{ days}$$
$$= 72 \text{ days}$$

Self-Check Answer
7.5%

Self-Check
If $26.25 interest was earned on $3,000 for 42 d, what was the rate of interest?

Problems

Solve the following problems for either the principal, the rate, or the time. Use the banker's year (360 d). Carry out all interim money amounts to four decimal places where necessary. Carry out percents to the nearest tenth of a percent. Round your final answer to the nearest cent.

17. If $8.58 interest was paid at 7% for 60 d, what was the amount of the loan?

18. What rate of interest was being charged if $25 interest was paid on $1,250 for 90 d?

19. For how many days was a $20,000 loan outstanding if it earned $500 interest at a rate of $5\frac{1}{2}\%$?

20. How much was borrowed if interest for 96 d came to $483.98 at a rate of $7\frac{1}{2}\%$?

21. If $5 interest was earned on $612.25 at a rate of $6\frac{3}{4}\%$, for how many days was the loan outstanding? Round the fraction of a year to four decimal places and your final answer to the nearest whole number.

22. What was the amount of a loan if $40.29 interest was earned for 42 d at a rate of 9%?

23. For how many days was a loan outstanding if $15.40 interest was paid on $1,050 at a rate of $8\frac{1}{2}\%$? Round the fraction of a year to four decimal places and your final answer to the nearest whole number.

24. What was the rate of interest if $50.31 interest was paid on $5,160 for 36 d?

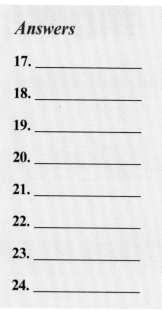

Answers

17._____

18._____

19._____

20._____

21._____

22._____

23._____

24._____

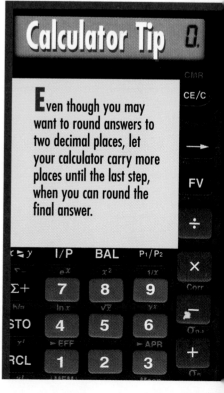

Calculator Tip

Even though you may want to round answers to two decimal places, let your calculator carry more places until the last step, when you can round the final answer.

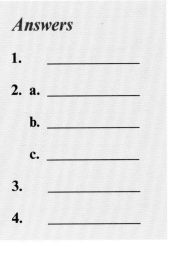

Business Applications 7.2

Answers

1. _____

2. a. _____

   b. _____

   c. _____

3. _____

4. _____

1. Tryon Dresses was billed $39 interest on a $1,500 payment that was 60 d late. What rate of interest was charged?

2. Juliet Jewelry charges 18% interest on delinquent accounts. One customer was 42 d late in making a payment on an account of $1,815. The customer sent a check for the cost of the merchandise, plus an interest payment of $32.67.

   a. Was $32.67 the correct amount of interest?

   b. If not, what was the correct amount?

   c. How many days did the interest payment the customer sent actually cover?

3. Kovel Antique Associates were charged $617.42 interest on a 90-d loan in the amount of $23,900. What rate of interest was Kovel charged?

4. You invest $750 in an account earning simple interest quarterly. The rate is also adjusted quarterly. You leave the money invested for 9 months. In the first two quarters the rate is 5% and in the third quarter the rate drops to 4%. Compute the interest earned over the 9 months.

Activities 7.2

Graph Activity

Discount Furniture Company borrowed $120,000 for six months to finance its inventory. The interest rate the bank charged was $8\frac{3}{4}\%$, simple interest. How much interest will Discount Furniture Company pay on this loan? _____

More discount buying

Total department store sales rose 6.4% in 1997, to $264.8 billion, and discount stores took a bigger bite of the market. Share by type of store and change from 1996:

Share
(change)

Discount
62.4%
(+1.5%)

22.0%
(-0.4%) ← Conventional

15.6%
(-1.2%)

National chain[1]

1 – Sears, JCPenney, etc.

Source: International Mass Retail Association

By Cindy Hall and Dave Merrill, USA TODAY

Challenge Activity

Megan wants to buy a canoe for $1,400. She has the option of paying for it with her Visa (the annual rate on her Visa is 15%), taking out a 12-month loan at the bank at 12% with monthly payments, or withdrawing the money from her savings account. Her savings account is earning 4%. Which method of payment will cost Megan the least amount of money and are there any reasons that she should not choose that method?

Internet Activity

What is the minimum amount of money you need to open a personal money market and get an interest rate of 2.8000%? What is the annual percentage yield? If you had the same amount of money and put it in a savings account, what would be the interest rate? What would be the annual percentage yield? Go to

http://www.charterbank.com,

choose **Products and Services** then click on **Savings and Investment Rates.**

2.800%

Name:_____ Date:_____

SKILLBUILDER 7.3

Computing Exact Interest and Exact Time

Learning Objectives

- **Compute exact interest.**
- **Use a table to compute exact time.**

Computing Exact Interest

The United States government always uses a 365-d year to compute interest. This is called **exact interest,** and it is more accurate than interest computed on a banker's year, since it is based on the true number of days in a year. To compute exact interest, use the simple interest formula $I = P \times R \times T$, but change the denominator of the time fraction from 360 to 365.

EXAMPLE

Compute the exact interest due on a loan of $3,500 at 12% for 103 d.

SOLUTION

$$I \qquad = \quad P \; \times \; R \; \times \; T$$
$$\$118.520 \rightarrow \$118.52 = \$3,500 \times 0.12 \times \frac{103}{365}$$

 Self-Check
Compute the exact interest due on a loan of $4,100 at 9% for 225 d.

 **Self-Check Answer**
$227.47

Problems

Answers

1. _____

2. _____

3. _____

4. _____

5. _____

6. _____

7. _____

8. _____

Compute the exact interest due on the following loans. Where necessary, count the precise number of days between dates.

1. $11,000 for 180 d at 6%

2. $942 from July 10 to September 16 at $8\frac{1}{2}$%

3. $9,867 from August 20 to December 7 at $5\frac{3}{4}$%

4. $1,550 from November 14 to April 14 at 9%

5. $15,000 from May 4 to August 10 at 8%

Student Success Hints

Taking Tests
Listen to and carefully follow directions. Ask appropriate questions.

6. $6,000 for 240 d at $8\frac{1}{2}$%

7. $1,286 for 77 d at $6\frac{1}{4}$%

8. $775 for 48 d at $7\frac{1}{2}$%

Using exact time and exact interest, find the missing element in each of the following problems.

	Principal	×	Rate	×	Time	=	Interest
9.	$7,000	×	____	×	90 d	=	$120.82
10.	$4,000	×	9%	×	____	=	313.64
11.	_____	×	7%	×	68 d	=	20.87
12.	$12,800	×	$5\frac{3}{4}\%$	×	____	=	350.86

Using a Table to Compute Exact Time

The Simple Interest Table in the Appendix can be used to count the exact number of days between two dates. Each day of the year is consecutively numbered, from 1 for January 1 to 365 for December 31. To use the table, follow these steps.

1. Select the column with the appropriate month heading.

2. Move down that column to the row that is the date of the day you are seeking. (Note that the extreme left and right columns are numbered consecutively 1 to 31.)

3. The intersection of the date row and month column is the consecutive number of that day in the year.

4. For the second date, repeat steps 1 to 3.

5. Subtract the smaller number from the larger. The difference is the exact number of days between the two dates.

The table assumes a 365-day year. If the loan period includes the month of February in a leap year, add one to the consecutive number of each day after February 28. February 29 would be day 60, March 1 would become day 61, and so on.

MATH TIP

Remember, the table assumes a 365-d year. If the loan period includes February in a leap year, add a day to the total number of days.

EXAMPLE

What is the exact number of days between November 28 and March 15?

SOLUTION

Go to the column headed "November," and move down to row 28. November 28 is day 332. Subtract 332 from 365 to count 33 days in the old year. March 15 is day 74.

$$74 + 33 = 107$$

The number of days is 107. (If this were a leap year, March 15 would be day 75, and the total would then be 108 days.)

Self-Check
Find the exact number of days between March 12 and December 12.

Self-Check Answer
275

Answers

13. _____

14. _____

15. _____

16. _____

17. _____

Determine the exact number of days between the following dates.

13. March 12 and August 23

14. September 8 and December 19

15. January 8 and May 19 (leap year)

16. December 27 and February 11

17. April 15 and October 21

Name:_____ Date:_____

18. November 25 and March 5

Determine the exact time and interest for the following loans, using the Table of Days Between Two Dates in the Appendix to count the days.

Amount of Loan	Rate	Term of Loan	Method	Number of Days	Interest Due
19. $ 3,590.00	8%	8/12 to 11/9	Banker's	_____	_____
20. 1,527.00	$7\frac{1}{2}$%	1/18 to 4/11 (nonleap year)	Exact	_____	_____
21. 450.00	$5\frac{1}{2}$%	12/14 to 2/28	Exact	_____	_____
22. 5,872.00	$6\frac{1}{4}$%	6/23 to 12/15	Banker's	_____	_____
23. 1,624.00	9%	4/26 to 6/12	Banker's	_____	_____
24. 8,527.00	$7\frac{3}{4}$%	1/16 to 4/10	Exact	_____	_____
25. 37,693.00	6%	10/9 to 2/6	Banker's	_____	_____

Answers

26. _____

27. _____

28. _____

29. _____

30. a. _____

b. _____

Use exact time and exact interest unless otherwise indicated.

26. A $16,000 loan at $6\frac{1}{2}$% was made on May 13 and was repaid on July 12. How much interest was paid?

27. A $5,000 loan was made on January 12. On April 12 a total amount of $5,093.75 was repaid. What rate of interest was charged?

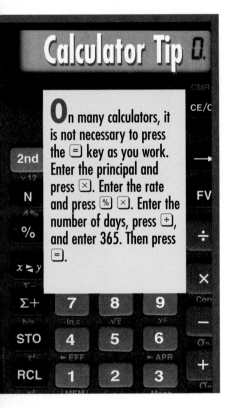

28. A $23,900 loan at $7\frac{3}{4}$% was charged $304.48 interest. For how long was the loan outstanding?

Calculator Tip

On many calculators, it is not necessary to press the ⊟ key as you work. Enter the principal and press ⊠. Enter the rate and press ⍰ ⊠. Enter the number of days, press ⊕, and enter 365. Then press ⊟.

29. $419.18 interest was paid on a 90-d loan at $8\frac{1}{2}$%. What was the amount of the loan to the nearest whole dollar?

30. A loan of $86,427 was made for 69 d at $5\frac{1}{2}$% interest. How much interest would be paid (a) using the banker's year method and (b) using the exact interest method?

Business Applications 7.3

1. The Last National Bank of Seattle loaned $375,000 to one of its best corporate customers. Interest was charged at one and a half points over prime for a period of 32 d. Prime rate is 6%. Using the exact interest basis, how much interest would they pay?

2. Michelle Carlson, treasurer of Juliet Jewelry, has discovered an error in Juliet's tax payments. The additional $13,000 due is subject to a penalty of 25% of the amount owed, plus exact interest at a rate of 9%, from April 15 to July 21, the date of the payment. What is the total amount that must be paid?

3. Shaydee Enterprises' tax return for the previous year was audited by the IRS. As a result of this audit on October 19, it was determined that they owed an additional $28,000 in tax. In addition to the tax payment, penalty interest is charged at 120% percent of the normal rate of 9% from April 15 to the audit date plus a further penalty of 5% per month, not to exceed 25%. What was the total amount of the payment due from them?

4. Pat Parker was to receive an $800 refund on his federal tax payment for the previous year. Because the refund was not made until June 11, he is entitled to exact interest at a rate of 8% from April 15 to the refund date. How much interest should he receive?

5. A commercial loan of $41,750 was obtained for the period of November 15 to April 15, with interest at $6\frac{7}{8}$% on an exact basis. What is the total amount to be repaid on April 15?

Activities 7.3

Graph Activity

On September 1, Jay Tefoe, a junior in high school, deposited $1,500 that he had earned from his summer job into an account at County Bank. This account paid simple interest at the rate of 5%. If the bank uses the exact interest method and credits Jay's account for all interest earned on December 31, what will be the balance in this account? _____

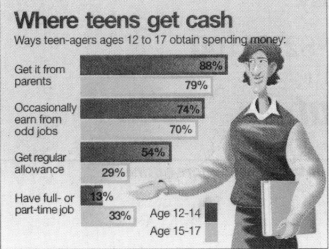

Where teens get cash

Ways teen-agers ages 12 to 17 obtain spending money:

	Age 12-14	Age 15-17
Get it from parents	88%	79%
Occasionally earn from odd jobs	74%	70%
Get regular allowance	54%	29%
Have full- or part-time job	13%	33%

Source: ICR TeenEXCEL survey for Merrill Lynch   By Anne R. Carey and Bob Laird, USA TODAY

Challenge Activity

Wayland Services is seeking a loan of $87,000 for 120 days. The Last National Bank of Seattle quoted a rate of 7.5% on an exact interest basis. Banco Bunko quoted a rate of 7.4% using the banker's basis. Which is the better offer? Explain. _____

Internet Activity

There are many different types of loans. Check this site and find out the terms of a seller-carry loan.

Loan
$7\frac{1}{4}\%$

http://homeadvisor.msn.com/highlight/qa/4996.as

SKILLBUILDER 7.4

Computing Compound Interest

Learning Objectives

- **Compute compound interest using a formula.**
- **Compute compound interest using a periodic rate.**
- **Compute compound interest using a table.**
- **Find interest compounded daily using a table.**

Using a Formula

When money is deposited in a savings bank or other savings institution, the account earns **compound interest.** This means that the interest made on an account is periodically combined with the principal so that the base amount on which interest is paid becomes greater and greater. To compute compound interest, follow these steps.

1. Find the simple interest on the principal for one interest period, using the simple interest formula $I = P \times R \times T$.

2. Add this interest to the first period principal to find the principal for the second period.

3. Compute the interest earned on the second period principal.

4. Add the second period interest to the second period principal to find the third period principal.

5. Repeat this procedure for each period for which compound interest is to be earned. The total of the original principal and all interest earned is the **compound amount.** To find the compound interest earned, subtract the original principal from the compound amount.

When interest is compounded annually, the value of T is 1. However, interest is usually compounded more than once a year. When this happens, replace the time factor, T, in the simple interest formula by a common fraction representing the compounding period (semiannual, $\frac{1}{2}$; quarterly, $\frac{1}{4}$; monthly, $\frac{1}{12}$; daily, $\frac{1}{365}$).

EXAMPLE

A deposit of $5,000 is made to a savings account earning interest at 4% compounded annually. How much would be in the account at the end of 2 y if the interest is left in the account? What is the compound interest?

SOLUTION

1. *Year 1:* $5,000 $\times$ 0.04 $\times$ 1 = $200 interest

2.   + 200   First year's interest

3. *Year 2:* $5,200 $\times$ 0.04 $\times$ 1 = $208 interest

4.   + 208   Second year's interest

           $5,408   Compound amount

5. $5,408   Compound amount

         − 5,000   Original principal

           $408   Compound interest

Simple interest would have resulted in only $400 interest ($200 for each year).

Self-Check

Find the interest if $15,000 is compounded monthly at 5% for 1 y.

Self-Check Answer

$767.41

Problems

Find (a) the compound amount and (b) the compound interest for these accounts. Round all money amounts to the nearest cent where necessary.

1. $1,500 at 4% compounded annually for 3 y

Answers

1. a. _____

   b. _____

2. $6,000 at 3% compounded annually for 4 y

3. $3,000 at $2\frac{1}{2}$% compounded semiannually for 4 y

4. $12,000 at 6% compounded quarterly for 2 y

5. $4,500 at 3% compounded bimonthly for 2 y

6. $8,000 at 4% compounded monthly for 1 y

Answers

6. a. _____

   b. _____

7. a. _____

   b. _____

7. $11,000 at $3\frac{1}{2}$% compounded every 4 months for 2 y

Using a Periodic Rate

Compound interest can be computed by using the rate of interest for a compounding period in the formula instead of the annual rate.

EXAMPLE

A deposit of $5,000 is made to a savings account earning interest at 4% compounded semiannually for 1 y. Find the compound amount and the compound interest.

SOLUTION

1. Determine the number of compounding periods in 1 y (semiannually = 2 periods a year).

2. Divide the annual rate of interest by the number of compounding periods in a year (4% ÷ 2 = 2%).

3. Add 1 to the period rate found in step 2 (0.02 + 1 = 1.02). When the principal on deposit is multiplied by a factor such as this, the product is the compound amount—principal plus interest—for that period.

4. Multiply the total time on deposit by the number of compounding periods in 1 y to determine the number of compounding computations to be made (1 y × 2 periods a year for semiannual compounding = 2 compounding periods).

5. Multiply the beginning principal by the factor obtained in step 3. Continue to multiply the resulting product by this factor for the required number of compounding periods obtained in step 4.

Period 1: $5,000 × 1.02 = $5,100
Period 2: $5,100 × 1.02 = $5,202   Compound amount
                   − 5,000   Original principal
                   $ 202   Compound interest

Self-Check Answers
$5,412.16
$412.16

Self-Check
Find (a) the compound amount and (b) the compound interest if $5,000 earns 4% interest and is compounded semiannually for 2 y.

Using the preceding method, find (a) the compound amount and (b) the compound interest for these accounts. Carry out the periodic rate to four decimal places where necessary.

8. $850 at 3% compounded semiannually for 2 y

9. $3,600 at 5% compounded every 3 months for 2 y

10. $7,500 at 6% compounded quarterly for $\frac{3}{4}$ y

11. $9,875 at $2\frac{1}{2}$% compounded bimonthly (every 2 months) for $\frac{2}{3}$ y

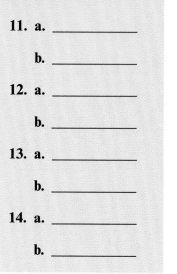
12. $1,500 at $4\frac{1}{2}$% compounded monthly for 8 mo

13. $12,000 at 3% compounded monthly for $\frac{1}{2}$ y

14. $1,850 at 4% compounded every 4 months for 1 y

Using the Compound Interest Table

The Compound Interest Table in the Appendix lists factors needed to compute compound interest at various rates for up to 25 interest-compounding periods.

EXAMPLE

Compute the compound amount and compound interest on $5,580 at 6% compounded quarterly for 5 y.

SOLUTION

1. Determine the number of compounding periods in 1 y (quarterly = 4).

2. Divide the annual rate of interest by the number of compounding periods in 1 y to get the interest rate per period (6% ÷ 4 = $1\frac{1}{2}$%).

3. Multiply the total number of years on deposit by the number of compounding periods in 1 y to get the total number of compounding periods (5 y × 4 = 20).

4. Locate the appropriate interest rate column in the Compound Interest Table in the Appendix ($1\frac{1}{2}$%). Move down the rate column to the line matching the total number of periods determined in step 3, in this case, 20. The factor at this point is the value of $1 compounded at that rate for that period of time: 1.346854.

5. Multiply the original principal by the compound factor to get the compound amount.

Compound Amount	=	*Original Principal*	×	*Compound Factor*
$7,515.45	=	$5,580	×	1.346854

$7,515.45 Compound amount
− 5,580.00 Original principal
$1,935.45 Compound interest

Self-Check
Determine (a) the compound interest and (b) the compound amount on $4,000 at 6% compounded bimonthly for 2 y.

Self-Check Answers
$507.30
$4,507.30

Use the Compound Interest Table in the Appendix to find the compound amount and the compound interest for the following accounts.

	Principal	Annual Int. Rate	Time on Deposit	Compounded	Compound Amount	Compound Interest
15.	$ 8,000	4%	4 y	Quarterly	_____	_____
16.	4,100	2%	12 y	Semiannually	_____	_____
17.	1,260	5%	6 y	Quarterly	_____	_____
18.	6,975	3%	$12\frac{1}{2}$ y	Semiannually	_____	_____
19.	7,500	2%	3 y	Quarterly	_____	_____
20.	12,000	$2\frac{1}{2}$%	2 y	Semiannually	_____	_____
21.	4,500	4%	2 y	Monthly	_____	_____
22.	5,000	$4\frac{1}{2}$%	3 y	Every 2 mo	_____	_____
23.	3,700	2%	1 y 8 mo	Every 4 mo	_____	_____
24.	4,000	4%	1 y	Monthly	_____	_____
25.	4,000	3%	$3\frac{1}{2}$ y	Every 2 mo	_____	_____
26.	38,500	$4\frac{1}{2}$%	3 y	Every 4 mo	_____	_____

Name:_____ Date:_____

Finding Interest Compounded Daily

The Compound Interest Table $1 Compounded Daily for a 365-Day Year in the Appendix is a compound interest table for daily compounding of interest. It is customary to credit the interest earned at the end of each month when interest is compounded daily. To use this table, select the appropriate annual rate column: then read down the column to the number of days for the compounding factor.

EXAMPLE

How much interest would $5,000 earn at a rate of 3% compounded daily for 30 d?

SOLUTION

Read down the 3% column to line 30. The factor is 1.0024659. Then multiply the principal by the compounding factor to find the compound amount.

$5,000 × 1.0024659 = $5,012.33 Compound amount

Subtract the original principal from the compound amount to find the compound interest.

$5,012.33 Compound amount
− 5,000.00 Original principal
$12.33 Compound interest

 Self-Check
How much would $2,500 earn at a rate of 3.5% compounded daily for 25 d?

 Self-Check Answer
$5.99

Use the Compound Interest Table $1 Compounded Daily for a 365-Day Year in the Appendix to find the compound amount and the compound interest earned by these accounts. Use seven decimal places for the compound factor.

	Principal	Annual Int. Rate	Time on Deposit	Compound Amount	Compound Interest
27.	$ 8,000	$2\frac{1}{2}\%$	23 d	_____	_____
28.	11,500	3%	31 d	_____	_____
29.	35,000	$3\frac{1}{2}\%$	5 d	_____	_____
30.	1,300	2%	29 d	_____	_____
31.	19,800	$2\frac{3}{4}\%$	19 d	_____	_____
32.	14,159	$3\frac{1}{2}\%$	12 d	_____	_____

Business Applications 7.4

Answers

1. _____

    _____

2. _____

    _____

3. _____

1. Weber Industries placed $15,000 in a savings account that paid 4% interest compounded monthly for 1 y in order to finance equipment purchases at the end of the year. Find the compound interest and the compound amount in the account at the end of the year.

2. In order to plan for an anticipated expansion, Lori's Boutique put $22,000 in an account in which interest was paid at 3% compounded quarterly. Find the compound interest and the compound amount at the end of 9 mo.

3. Using the periodic rate method described on page 408, compute the compound interest on $11,500 at 7%, compounded monthly for 9 months.

4. Compare the compound interest earned on a $25,000 6 month certificate of deposit earning 6%, compounded monthly and semi-monthly.

5. A $50,000 bond was posted on November 21, 1999 to guarantee completion of a construction project. The bond will earn 8% compounded monthly. The project was begun on November 21, 1999, and was completed 21 months later. What was the completion date of the project, and the value of the bond at that time?

6. How much interest would $25,000 earn in 10 days at a rate of 5% compounded daily?

Activities 7.4

Graph Activity

Murphy Sanders put $10,000 in an investment to supplement her retirement. The investment will earn 14%, compounded semi-annually. Calculate the value of this investment at the end of each year for the first 5 years.

Year 1 _____

Year 2 _____

Year 3 _____

Year 4 _____

Year 5 _____

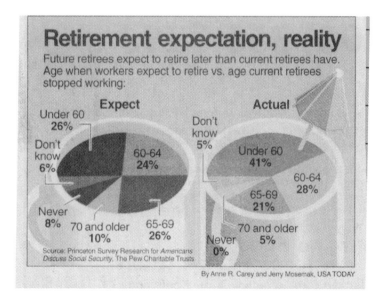

Retirement expectation, reality

Future retirees expect to retire later than current retirees have. Age when workers expect to retire vs. age current retirees stopped working:

Expect

Under 60
26%

Don't know
6%

60-64
24%

Never
8%

70 and older
10%

65-69
26%

Actual

Don't know
5%

Under 60
41%

60-64
28%

65-69
21%

70 and older
5%

Never
0%

Source: Princeton Survey Research for *Americans Discuss Social Security*, The Pew Charitable Trusts

By Anne R. Carey and Jerry Mosemak, USA TODAY

Challenge Activity

Which investment provides the best return, $8,000 at 2% compounded quarterly for 4 years, or $8,000 at 2% compounded every 4 months? Explain.

Answers

_____

_____

Internet Activity

Would you work for a penny a day? A consultant asked for a salary of a penny a day as a salary if the salary doubled every day. Her employer agreed, but wanted to fire her at the end of the month because he couldn't afford to pay her. Find out what her salary was on the 31st of the month.

http://www.russell.com/services/in.../employee/articles/scarticle2b.html

SKILLBUILDER 7.5

Discounting Noninterest-Bearing Notes and Drafts

Learning Objectives

* **Find the proceeds of a noninterest-bearing note.**
* **Find the proceeds of a noninterest-bearing draft.**

Finding the Proceeds of a Noninterest-Bearing Note

A **promissory note** is a written promise to pay a sum of money at a specific time. The date on which the payment is promised is the **due date,** or **maturity date.** The maturity date of the note may be specified exactly, or it may be stated as a definite number of days, months, or years from the date of issue of the note. The time period from the date of issue to the maturity date is the **term of the note.**

To find the maturity date of a 60-day note dated July 20, count the exact number of days from the date of issue to determine the date on which the note is to be paid. If the term of the note is stated in months, count forward from the date of the note the given number of months to the same date in the month of maturity. For example, a 2-mo note issued July 15 would be due September 15. If cash is needed before the maturity date of the note, the business may sell the note to a bank. This process is called **discounting the note,** and the date of sale to the bank is the **discount date.** The **term of discount** is the exact number of days from the discount date to the maturity date.

When a note is discounted, the bank charges banker's interest, the **amount of discount.** The interest is computed on the face value for the term of discount and is then deducted from the face value of the note. The balance, called the **proceeds** of the note, is paid to the borrower. A typical noninterest-bearing note is shown below.

$ *1,500.00*	*April 17,* 20 ___	
___*Sixty (60) days*___ after date ___*J*___	Promise to pay to	
the order of *Kris Koshak*		
One thousand five hundred and ⁰⁰/100 ——————— Dollars		
Payable at *Morgan State Bank*		
Value recieved *with no interest*		
No. __*1785*__ Due *June 16,* 20 ___	*John Perez*	

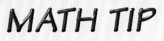
EXAMPLE

Find the maturity date and term of discount of a 90-d note dated June 20 and discounted July 9.

SOLUTION

Use the Table of Days Between Two Dates in the Appendix to determine the exact number of days between the two dates.

> June 20 = 171 + 90 = 261
> 261 = September 18   **Maturity date**
>
> July 9 = 190
> 261 − 190 = 71 d   **Term of discount**

If the same note ran for 3 mo and was discounted on the same date, the maturity date would be September 20 and the term of discount from July 9 to September 20 would be 73 d.

EXAMPLE

If the note on page 419, signed by John Perez, was discounted on May 1 at 9%, find the proceeds.

SOLUTION

$1,500.00	Face value of note
	Maturity date: April 17 = 107 + 60 days = 167 = June 16
	Term of discount: May 1 = 121; 167 − 121 = 46 d
− 17.25	Less amount of discount: $1,500 × 0.09 × $\dfrac{46}{360}$
$1,482.75	Proceeds

Self-Check Answers
Apr 1
72 d
$198
$10,802

Self-Check
Find (a) the maturity date, (b) the term of discount, (c) the amount of discount, and (d) the proceeds for an $11,000 note if the term of the note is 120 d, the date is December 2, the discount date is January 19, and the rate of discount is 9%.

Problems

1. Refer to the promissory note on page 419. From this note, complete the following information:

   a. Principal _____ b. Term of note _____

   c. Date of Issue _____ d. Payee _____

   e. Rate _____ f. Maker _____

   g. Maturity date _____

2. Find the maturity date and term of discount for each of the following:

Date of Note	Term of Note	Maturity Date	Discount Date	Term of Discount
a. September 30	30 d	_____	October 15	_____
b. January 18, 20—	45 d	_____	January 31	_____
c. March 31	3 mo	_____	May 4	_____
d. September 25	120 d	_____	October 25	_____
e. May 12	72 d	_____	June 15	_____
f. November 26	210 d	_____	March 26	_____
g. June 4	60 d	_____	July 5	_____
h. February 29	90 d	_____	April 1	_____
i. August 31	6 mo	_____	December 30	_____
j. October 1	54 d	_____	October 31	_____

3. Find the maturity date, term of discount, amount of discount, and proceeds for the following noninterest-bearing notes. Use the banker's year.

	Face Value	Term of Note	Date of Note	Discount Date	Rate of Discount	Maturity Date	Term of Discount	Amount of Discount	Proceeds
a.	$500	30 d	Jan. 5	Jan. 15	9%	_____	_____	_____	_____
b.	1,750	90 d	Mar. 26	May 2	11%	_____	_____	_____	_____
c.	5,000	2 mo	Nov. 18	Dec. 31	$7\frac{1}{2}\%$	_____	_____	_____	_____
d.	8,659	72 d	June 12	July 1	10%	_____	_____	_____	_____
e.	2,500	45 d	Sept. 8	Sept. 30	$6\frac{3}{4}\%$	_____	_____	_____	_____
f.	985	60 d	Oct. 15	Nov. 1	8%	_____	_____	_____	_____
g.	7,850	210 d	May 20	Aug. 26	$11\frac{1}{4}\%$	_____	_____	_____	_____
h.	12,500	4 mo	Aug. 31	Nov. 1	$9\frac{1}{2}\%$	_____	_____	_____	_____
i.	6,250	180 d	Feb 12, 20—	July 10	$10\frac{5}{8}\%$	_____	_____	_____	_____

Finding the Proceeds of a Noninterest-Bearing Draft

A **draft** is a written order for payment made by one party to another. It differs from a check in that it must be accepted before it becomes legally binding. If the draft calls for payment at a specified date, it is a **time draft,** in which case the date of acceptance is not important. When the draft calls for payment a certain time after sight, or acceptance, it is a **sight draft,** and the maturity date can be determined only after the draft has been presented for acceptance. The parties to a draft are the **drawer,** the party creating the draft; the **payee,** the party to whom payment is to be made; and the **drawee,** the party being ordered to pay. With a sight draft, the drawee is also the **acceptor.**

$ _1,500.00_	_April 17,_ 20 ___
Sixty (60) days from date	Pay to the
order of _Kris Koshak_	
One thousand five hundred and 00/100	Dollars
Value recieved and charge to the account of	
To _Last National Bank of Secaucus)_	
No. _782_	/s/ John Bailey

A draft is discounted in the same way as a promissory note.

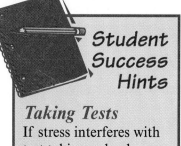

Student Success Hints

Taking Tests
If stress interferes with test taking, relax by closing your eyes and taking a couple of deep breaths.

EXAMPLE

Find the proceeds on a sight draft for $3,850. Its maturity date is 45 d after sight. It was accepted on November 15 and discounted on December 15 at 9%.

SOLUTION

$3,850.00 Face value of draft
Maturity date: 45 d from November 15 = December 30
Term of discount: December 15 to December 30 = 15 d

– 14.44   Less amount of discount: $3,850 $\times$ 0.09 $\times \dfrac{15}{360}$

$3,835.56 Proceeds

**Self-Check
Answers**
Oct 4
84 d
$178.50
$8,821.50

Self-Check

Find (a) the maturity date, (b) the term of discount, (c) the amount of discount, and (d) the proceeds for a $9,000 noninterest-bearing draft if the date of the draft is June 4, the due date is 4 mo from the date, the acceptance date is July 12, the discount date is July 12, and the rate of discount is $8\frac{1}{2}\%$. Use the banker's year.

4. From the information provided on the time draft shown on page 423, identify the following:

a. Principal _____   b. Term of draft_____

c. Date of issue _____   d. Drawer _____

e. Payee _____   f. Drawee

g. Maturity date _____

_____

Calculator Tip

You can enter percents in decimal form or use the whole-number percent and the percent key.

5. Find the maturity date, term of discount, amount of discount, and proceeds for the following noninterest-bearing drafts. Use the banker's year. Use the space on page 426 to work out your answers.

	Face Value	Date of Draft	Due Date	Acceptance Date	Discount Date	Rate of Discount	Maturity Date	Term of Discount	Amount of Discount	Proceeds
a.	$ 2,000	March 30	60 d from date	—	April 30	$8\frac{1}{2}\%$				
b.	5,500	September 12	45 d after sight	October 2	October 15	10%				
c.	1,200	February 15	3 mo from date	—	March 1	6%				
d.	875	October 9	90 d after sight	November 1	November 2	9%				
e.	3,750	May 17	72 d from date	—	June 1	$7\frac{1}{4}\%$				
f.	11,200	December 11	15 d after acceptance	February 8	February 10	$5\frac{1}{8}\%$				
g.	7,683	August 26	45 d from date	—	September 12	8%				
h.	1,150	April 5	10 d after sight	June 3	June 3	$6\frac{3}{4}\%$				
i.	2,900	November 28	30 d from date	—	December 1	$9\frac{1}{2}\%$				

Use the space below to solve Problem 5.

Business Applications 7.5

1. On May 24, Melinda Phelps was having cash-flow problems and was unable to pay the $7,800 she owed for the latest shipment of merchandise purchased from High Seas Imports. Since she is an excellent customer of High Seas Imports, they have agreed to extend credit to her for 72 d at no interest. However, Melinda must sign a promissory note for the credit extension. On maturity, the promissory note is to be paid into High Seas' account at the Last National Bank of Secaucus. On June 28, High Seas discounted the note at 9%. From the information provided, complete the following promissory note, and compute the proceeds.

$ _____   _____ 20 _____

_____ after date _____ promise to pay to

the order of _____

_____ Dollars

Payable at _____

Value recieved _____

No. _____ Due _____ 20 _____   _____

2. On September 11, the buyer for High Seas Imports arranged the purchase of $112,000 of merchandise from Zodiac Products. A 120-d time draft was given to Zodiac in payment for the shipment through High Seas' account at Buck's Bank International. On November 11, Zodiac discounted the draft with its bank at $8\frac{3}{4}\%$ interest. From the information provided, complete the following draft and compute the proceeds that Zodiac received from its bank.

$ _____   _____ 20 _____

_____ *Pay to the*

order of _____

_____ *Dollars*

Value recieved and charge
to the account of

To _____

No. _____   _____   _____

3. You are self-employed and one of your suppliers took a non-interest bearing note from you. Is the lack of interest good or bad for your taxes?

Activities 7.5

Graph Activity

An airline borrowed some money from a bank to provide additional working capital necessary to cover the cost of a strike. The face amount of the 3-month note was $580,000 and the discount rate 12%. Calculate the proceeds. _____

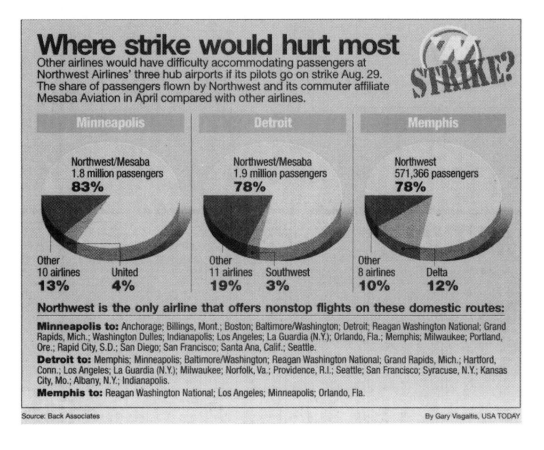

Where strike would hurt most

Other airlines would have difficulty accommodating passengers at Northwest Airlines' three hub airports if its pilots go on strike Aug. 29. The share of passengers flown by Northwest and its commuter affiliate Mesaba Aviation in April compared with other airlines.

STRIKE?

Minneapolis

Northwest/Mesaba
1.8 million passengers
83%

Other
10 airlines
13%

United
4%

Detroit

Northwest/Mesaba
1.9 million passengers
78%

Other
11 airlines
19%

Southwest
3%

Memphis

Northwest
571,366 passengers
78%

Other
8 airlines
10%

Delta
12%

Northwest is the only airline that offers nonstop flights on these domestic routes:

Minneapolis to: Anchorage; Billings, Mont.; Boston; Baltimore/Washington; Detroit; Reagan Washington National; Grand Rapids, Mich.; Washington Dulles; Indianapolis; Los Angeles; La Guardia (N.Y.); Orlando, Fla.; Memphis; Milwaukee; Portland, Ore.; Rapid City, S.D.; San Diego; San Francisco; Santa Ana, Calif.; Seattle.

Detroit to: Memphis; Minneapolis; Baltimore/Washington; Reagan Washington National; Grand Rapids, Mich.; Hartford, Conn.; Los Angeles; La Guardia (N.Y.); Milwaukee; Norfolk, Va.; Providence, R.I.; Seattle; San Francisco; Syracuse, N.Y.; Kansas City, Mo.; Albany, N.Y.; Indianapolis.

Memphis to: Reagan Washington National; Los Angeles; Minneapolis; Orlando, Fla.

Source: Back Associates

By Gary Visgaitis, USA TODAY

Challenge Activity

Find out about the financial aid that is available at your school. Ask if there are various options to the type of loans available. Find out if there are certain criteria that need to be met to qualify for financial aid.

Internet Activity

What should you know before you sign
a promissory note for a loan?

http://faoman.ucdavis.edu/loaninfo.htm#befor

SKILLBUILDER 7.6

Discounting Interest-Bearing Notes and Drafts

Learning Objective

• **Find the proceeds on interest-bearing notes and drafts.**

Finding the Proceeds on Interest-Bearing Notes

Promissory notes and drafts may or may not bear interest. When a borrower uses any of these as **collateral,** that is, security for a loan, the bank pays the proceeds to the borrower and collects the maturity value when the obligation is due. The **maturity value** of an interest-bearing note or draft is the face value plus interest earned for the period of the note or draft. The bank lends the maturity value less the amount of discount (interest) computed on the maturity value. The borrower receives the remainder, or **proceeds.**

When a note or draft is collected by a bank, some banks deduct a certain percent of the maturity value of the note as a **collection fee.**

EXAMPLE

A $2,000, 90-d promissory note, dated July 2, with interest at 9%, is discounted on August 15 at $10\frac{1}{2}$%. Find the proceeds.

SOLUTION

Maturity date: 90 d from July 2 = September 30

$2,000.00	Face value
+ 45.00	Plus interest: $2,000 × 0.09 × $\frac{90}{360}$
$2,045.00	Maturity value
	Term of discount: August 15 to September 30 = 46 days
− 27.44	Less amount of discount: $2,045 × 0.105 × $\frac{46}{360}$
$2,017.56	Proceeds

EXAMPLE

A 30-d, $7\frac{1}{2}$% note with a face amount of $2,000 is discounted after 10 d at a rate of $8\frac{3}{4}$%. The bank charges a collection fee of 0.1% (0.001). Find the proceeds.

SOLUTION

$2,000.00	Face value
+ 12.50	Plus interest: $2,000 × 0.075 × $\frac{30}{360}$
$2,012.50	Maturity value
− 2.01	Less collection fee: $2,015.83 × 0.001
$2,010.49	
− 9.78	Less discount: $2,012.50 × 0.0875 × $\frac{20}{360}$
$2,000.71	Proceeds

Self-Check Answer
$5,939.87

Self-Check
A $5,784, 120-d promissory note, dated October 31, with interest at 11%, is discounted on February 1 at $12\frac{1}{2}$%. Find the proceeds.

Problems

1. Find the proceeds for each of the following notes. Use the banker's year.

	a.	b.	c.	d.
Face value	$1,000	$1,920	$516	$296.40
Term of note	2 mo	60 d	3 mo	90 d
Rate of interest	$8\frac{1}{2}\%$	$10\frac{1}{2}\%$	9%	$7\frac{1}{4}\%$
Amount of interest	$_____	$_____	$_____	$_____
Maturity value	$_____	$_____	$_____	$_____
Date of note	November 5	June 19	May 30	February 21, 2001
Discount date	December 6	July 9	July 16	April 4
Rate of discount	10%	$11\frac{1}{2}\%$	$10\frac{3}{4}\%$	9%
Maturity date	_____	_____	_____	_____
Term of discount	_____ d	_____ d	_____ d	_____ d
Amount of discount	$_____	$_____	$_____	$_____
Proceeds	$_____	$_____	$_____	$_____

Answer

2. _____

2. Lisa Hahn took a 120-d note for $1,500 from Sidney Wilde. The note was dated May 3, with interest at 9%. On May 18, Lisa Hahn sold the note to First Federal Bank at a discount rate of $10\frac{1}{4}\%$. What were the proceeds on the note?

3. Find the proceeds for each of these drafts, using the banker's year.

	a.	**b.**	**c.**	**d.**	**e.**
Face value	$2,250	$3,840	$6,000	$1,769	$2,750
Date of draft	May 5	July 12	August 8	October 19	January 17
When due	75 d from date	2 mo after sight	3 mo from date	45 d after sight	180 d from date
Acceptance date	May 12	July 21	August 12	November 1	March 17
Rate of interest	$6\frac{3}{4}\%$	0%	$7\frac{1}{4}\%$	0%	9%
Amount of interest	$ ____	$ ____	$ ____	$ ____	$ ____
Maturity value	$ ____	$ ____	$ ____	$ ____	$ ____
Maturity date	____	____	____	____	____
Discount date	May 15	August 4	August 15	November 8	May 15
Rate of discount	9%	$7\frac{3}{4}\%$	10%	$8\frac{1}{2}\%$	11%
Term of discount	____ d	____ d	____ d	____ d	____ d
Amount of discount	$ ____	$ ____	$ ____	$ ____	$ ____
Proceeds	$ ____	$ ____	$ ____	$ ____	$ ____

434   Unit 7   Interest and Discounts

© Glencoe/McGraw-Hill

4. Melody Manufacturers was given a 120-d, $8\frac{1}{2}$% note for $2,315. The note was dated July 14 and was discounted on September 1 at a rate of 11%. Find the proceeds.

5. Steve Zoltan held a 3-mo, 7% note for $750, dated March 31. He discounted the note at his bank on April 15 at a rate of $9\frac{1}{2}$%. The bank charges a collection fee of $2 or 0.1 percent, whichever is greater. What proceeds did Steve receive?

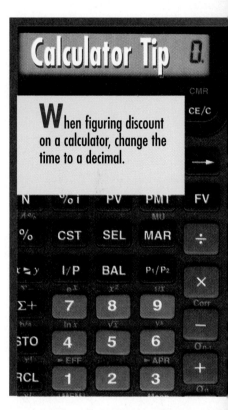

Calculator Tip

When figuring discount on a calculator, change the time to a decimal.

Business Applications 7.6

1. On March 30, Green Thumb Lawn Services purchased a new power sprayer for $3,875. The vendor extended credit to Green Thumb by taking a 135-d note at $7\frac{1}{2}\%$. On June 1, the vendor discounted the note at its bank at $9\frac{1}{4}\%$. The bank charges a collection fee of $2.50 or 0.1 percent, whichever is greater. What were the proceeds of the note?

2. On March 17, DaimlerChrysler AG issued a 45-d sight draft in the amount of $215,000 with interest at $6\frac{3}{4}\%$ exact interest from the date of issue to the maturity date. The draft was accepted on June 11 and discounted the same day at $7\frac{1}{2}\%$ banker's interest. Compute the amount of proceeds the payee will receive.

Student Success Hints

Taking Tests
If you have time, go back over the test and double check your work.

3. Isaac Freeman is buying a new compressor priced at $1,652. Air Products will finance the purchase but compute the interest on an exact basis. Isaac's bank will provide the same financing with interest computed on the banker's basis. If Isaac signs a 72-d 9% note on November 11, will the bank or Air Products provide the lower financing cost? If Air Products discounted the note at $10\frac{1}{4}\%$ on December 1, what proceeds will it receive?

Activities 7.6

Graph Activity

If you were a lending officer at a bank, would you lend money to these players? Why or why not?

Young draftees

This year's NBA draft pool includes four high school players. Previous high schoolers who were lottery picks and their rookie-season averages:

	Points	Rebounds
Kevin Garnett, Timberwolves (1995-96)	10.4	6.3
Kobe Bryant, Lakers (1996-97)	7.6	1.9
Tracy McGrady, Raptors (1997-98)	7.0	4.2

Source: USA TODAY research

By Scott Boeck and Bob Laird, USA TODAY

Challenge Activity

Explain in your own words how to find the proceeds on a promissory note with a discount.

Internet Activity

Collateral is divided into two broad classifications: tangible property and intangible property. Go to

http://www.mgovg.com/banking/ch3.html

and list nine types of collateral.

SKILLBUILDER 7.7

Installment Purchases

Learning Objectives

- **Compute the cost of buying on time.**
- **Compute the real rate of interest on an installment purchase.**

Computing the Cost of Buying on Time

Installment purchasing, or **buying on time,** is paying for an item by making a down payment and then making a series of payments, or installments, over a specified period of time. The usual practice is to charge interest for the privilege of making an installment purchase, although sometimes an installment purchase may cost no more than a cash purchase. The installment purchaser is generally charged various fees, such as an application fee for the loan, a credit investigation fee, and an insurance fee on the amount of the loan. These are added costs of buying on time.

To find the added cost of making an installment purchase, follow these steps.

1. Multiply the number of payments by the amount of each payment.

2. Add this product plus any fees or charges to the down payment. This total is the **total installment price.**

3. Subtract the cash purchase price from the total installment price.

The difference will be the added cost, if any, of buying on time.

EXAMPLE

Price Co. sells a color television for $450 cash. If the TV is purchased on an installment plan, a down payment of $45 and 15 monthly payments of $30.26 each are required. What is the added cost, if any, of buying the television on time?

SOLUTION

$453.90	Monthly payments: 15 × $30.26
+ 45.00	Plus down payment
$498.90	Total installment price
− 450.00	Less cash price
$48.90	Added cost of buying on time

Self-Check

The cash price of a copier is $1,300. If the copier is purchased on the installment plan, a down payment of $250 and 12 monthly payments of $104.60 are required. In addition, $10 is charged for a credit report. What is the added cost of buying on time?

Self-Check Answer
$215.20

Problems

1.

Item	Installment Terms			Installment Price	Cash Price	Cost of Buying on Time
	Down Payment	**Number of Payments**	**Monthly Payment**			
a. Word processor	None	12	$21.89	_____	$240.00	_____
b. CD player	$11.95	4	27.00	_____	119.95	_____
c. Stereo	None	15	20.77	_____	279.95	_____
d. Microwave	39.95	24	16.72	_____	379.95	_____
e. Cell phone	19.99	9	18.17	_____	169.99	_____
f. Refrigerator	58.00	30	21.32	_____	579.95	_____

2. George Manley purchased a $2,320 dishwasher for his café on an installment purchase plan. The purchase plan required a down payment of $232 and weekly payments of $30.75 for 78 wk. What was the added cost of buying on time?

Answers

2. _____

3. _____

3. Lauren Black purchased a used car priced at $8,795 for Lauren's Catering Service. In order to purchase on the installment plan, she had to make a 20% down payment and pay the balance in 24 monthly payments of $358 each. How much more did it cost to buy the car on time?

4. Rohm's Furniture purchased a ZIP486D computer for $2,799. It paid 15% down and financed the balance over 24 mo at $123 per month. How much did it cost to finance the purchase?

5. The Office Company purchased a 17,000-BTU air conditioner for a cash price of $697. It decided to finance the purchase with no money down and 12 monthly payments of only $65 per month. How much did it cost to finance the purchase?

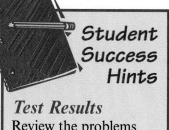

Student Success Hints

Test Results
Review the problems you did correctly to find your strong areas, good problem-solving techniques you used, the problems that were easiest for you to solve, and what short cuts or math tips that you used.

6. Ramon and Anita purchased five pieces of office furniture for their accounting business for $899, to be paid off with 18 payments of $57 per month. The payments include a $15 application fee and $10 for a credit report. How much did it cost to buy the furniture on time?

7. Sam's Pizza bought a new car priced at $7,304. After paying $1,000 down, the monthly payments were $125 per month for 60 mo. What was the total financing charge for this car?

8. Chamar's Carpets originally priced a 9- × 12-foot Persian carpet at $3,290. It was sale-priced at 40% off; Sunelda bought it by making 15 payments of $152 per month. How much more did Sunelda pay by buying the carpet on time?

9. Tom Meyer's Appliances originally sold a 30-in. stereo TV at the regular price of $799. Sue-Ellen bargained the price down to $699.95. Her monthly payments were $52 per month for 15 mo. How much more did it cost her to finance the purchase?

10. Roger's Copies leased a copier for 36 mo. The lease agreement called for a payment of $125 per month for 36 mo after a $500 down payment. Each payment contained interest and other charges, which came to $23.02 per month. What was the original price of the copier?

Computing the Real Rate of Interest

Businesses that finance purchases need to be aware of the Truth-in-Lending Act. This act applies to credit transactions involving personal, family, or household loans not exceeding $25,000. The act requires that the lender state the **true annual rate of interest (APR)** and the **finance charge,** which is any additional amount paid as a result of buying on credit instead of paying cash. The rate may be stated as "$1\frac{1}{2}$% a month (18% annual rate)." This tells the borrower that a rate of 18 percent is being charged only on the amount of money actually owed.

Add-on interest assumes that the borrower owes the entire sum of money over the period of the loan because the interest charged is computed for the entire period of the loan. Since a portion of the principal is being repaid with each payment, the borrower actually has the full use of the money for only a part of the period of the loan.

Neither add-on interest nor additional charges give a true picture of the real rate of interest being paid. To find the real rate of interest for a purchase, use this formula.

$$\text{Real rate of interest} = \frac{2 \times 12^* \times \text{added cost of buying on time}}{\text{amount of loan} \times (\text{number of payments} + 1)}$$

*If payments are made monthly, this number is 12. If payments are made weekly, this number is 52.

EXAMPLE

A CD player with a cash purchase price of $240 is sold on the installment purchase plan with no down payment. Add-on interest at an annual rate of 10% is charged to the purchaser, along with an application fee of $2.20, a credit check fee of $5, and insurance at a flat rate of $\frac{1}{2}$% of the amount of the loan. If the typewriter is to be paid for in 1 y, what is the cost of buying on time, and what is the monthly payment? Find the real rate of interest.

SOLUTION

$240.00	Cash price and amount of loan
24.00	Plus interest: $240 × 0.10 × 1
7.20	Plus application fee and credit check
+ 1.20	Plus insurance charge: $240 × 0.005
$272.40	Total installment price
− 240.00	Less cash price
$32.40	Added cost of buying on time
22.70	Monthly payment: $272.40 ÷ 12

$$\text{Real rate of interest} = \frac{2 \times 12 \times \text{added cost of buying on time}}{\text{amount of loan} \times (\text{number of payments} + 1)}$$

$$= \frac{2 \times 12 \times \$32.40}{\$240 \times (12 + 1)} = \frac{24 \times \$32.40}{\$240 \times 13}$$

$$= \frac{\$777.60}{\$3,120} = 0.2492 \text{ or } 24.9\%$$

**Self-Check
Answer**
17.3%

Self-Check

An office microwave was purchased with $39.95 down for 24 mo. The cash price was $379.95, and the monthly payment was $16.72. Find the real rate of interest.

Problems

11.

Item	Installment Terms				Real Rate of Interest
	Cash Price	Down Payment	Number of Payments	Monthly Payment	
a. Cell phone	$169.99	$19.99	9	$18.17	_____
b. File cabinet	76.50	6.50	12	6.50	_____
c. Stereo	279.95	None	15	20.77	_____

Answers

12. _____

13. _____

14. _____

12. In Problem 4, Rohm's Furniture purchased a computer for $2,799, paid 15% down, and financed the balance over 24 mo, with payments of $123 per month. What was the real rate of interest paid?

13. Midland Publishing leased a copier. The lease cost $828.75 to finance $3,671.28 over 36 months. What real rate of interest did Midland pay on the lease?

Calculator Tip

You can calculate the real rate of interest in one series of steps by multiplying the three factors in the numerator, dividing by the amount of the loan, and then dividing by the number of payments plus 1.

14. Tanya's Leather Goods sells a leather jacket for $285. It can be purchased for $51 per month over a 6-mo period. What real rate of interest is being paid?

15. Eduardo's Lawn Stuff sells a lawn tractor for $900. It can be purchased for $19 per month over a 5-y period. What real rate of interest is charged?

16. Pets and More purchased a Teensie notebook computer with a 2 gigabyte hard drive that sells for $1,399. The monthly payment for a 15-mo loan is $107.80 a month. (a) What is the cost of financing this purchase, and (b) what is the real rate of interest being paid?

17. "Gentleman Jim" Carvil is offering personal loans at a bargain rate of 3%, deducted in advance. His advertised special for a loan of $1,000 for one year requires a monthly payment of only $97. What is the real rate of interest being charged?

Business Applications 7.7

1. Lonnie Gwynere borrowed $5,000 for 4 months in order to purchase inventory for her auto parts shop. The bank charged 9% interest for the period on a banker's basis. She is required to pay back principal plus interest in four equal monthly installments. (a) How much is each installment? (b) Is she really paying 9% interest?

2. Elizabeth purchased an electronic surveyor's transit for $12,875. She was required to pay 20% down, 5% add-on interest for 2 y, a loan application fee of $25, a credit-check fee of $10, and insurance on the purchase at a flat rate of $\frac{1}{2}$% of the loan. Compute the amount of Elizabeth's monthly payment to the nearest dollar.

3. Pali's Department store purchased a copy machine. (a) Compute the amount of the monthly payment of the copier if it sells at $4,000 with a 10% down payment. Interest at 11% is $313.50; insurance is $60; credit report and application fee is $35; and a mandatory service contract is $10 per month. The period of the loan is 18 mo. (b) What is the real rate of interest being paid?

4. Michele Quan purchased a new pick-up truck for $17,500. She made a $4,000 down payment, and will pay the balance over 60 months at $290 a month. (a) What is the amount of the loan? (b) The total finance charge? (c) The real interest rate charged?

5. "Cap'n Bob" Edwards bought a new boat from Carvil's Crafts for $14,500 less a down payment of $2,500. The dealer arranged financing at a rate of 9.2% with 48 monthly payments of $342 each. Was "Gentleman Jim" correct in his rate quote of 9.2%?

6. A 2-year old leased vehicle is priced at $7,995 with a 10% down payment, to be financed at 10.99% over 5 years. The total finance charge is $2,229.90. Is the 10.99% interest rate a correct statement?

Activities 7.7

Graph Activity

Country Club Ford sells a Taurus to Rita Guthrie for $18,500 cash. If the car is purchased on an installment plan, a down payment of $1,500 and 36 monthly payments of $495 each are required. What is the added cost of buying the car on time? _____

Cost of keeping it running

Fuel costs have dropped in the past decade as a percentage of each dollar it costs[1] to drive a car, but maintenance and tire costs are up:

	1988	1998
Depreciation/interest	54.4%	54.0%
Fuel	18.1%	14.7%
Insurance	15.0%	15.2%
Maintenance	6.2%	8.0%
Tires	2.4%	3.4%
License	0.3%	0.6%
Other	3.5%	4.1%

1 – Costs for average owner to own and operate typical intermediate-size vehicle.   Source: Runzheimer International

By Cindy Hall and Marcy E. Mullins, USA TODAY

Challenge Activity

Define APR. Then give an example. What advice would you give to someone applying for a credit card or loan?

Internet Activity

Do you know the 11 credit card secrets banks do not want you to know? Find out what they are at **http://www.consumer.com/ consumer/CREDTIC.html**

SKILLBUILDER 7.8

Computing Interest on Unpaid Balances

Learning Objectives

- **Compute the installment purchase price on a month-to-month basis.**
- **Compute interest on the unpaid balance of an installment purchase.**

Computing the Installment Purchase Price

Because each payment made on an installment purchase pays back a portion of the principal, the fairest basis is to base the interest on the unpaid balance. This can be done by computing the interest on the balance each month and then subtracting the payment to find the new balance.

The usual practice is for equal installment payments to be made over the period of the loan. To find the amount of each installment, the principal plus total interest is divided by the number of payments.

EXAMPLE

A lawn mower was purchased on April 1 for $240 less a down payment of $60. The balance of $180 is to be paid off in equal monthly installments over a period of 3 mo with interest on the unpaid balance at 9%.

a. Prepare a table of the installment payments and balance due each month.

b. Find the amount of each payment if the loan is repaid in equal installments.

SOLUTION

a.

$$\text{Payment Toward Principal} = \text{Balance} \div \text{Number of Months}$$
$$\$60 = \$180 \div 3$$

$$\text{Interest Rate for Month} = \frac{\text{Interest Rate for Year}}{12}$$
$$= \frac{0.09}{12}$$
$$0.0075$$

$$\text{Interest for Month} = \text{Unpaid Balance} \times \text{Interest Rate for Month}$$
$$\$1.35 = \$180 \times 0.0075$$

Payment Date	Unpaid Balance	Interest for Month	Amount Due	Installment Payment	Balance Due
May 1	$180	$1.35	$181.35	$61.35	$120
June 1	120	0.90	120.90	60.90	60
July 1	60	0.45	60.45	60.45	0

b.
$180.00   Principal payments: 3 × $60
   2.70   Plus interest owed: $1.35 + $0.90 + $0.45

$182.70   Total

$$\text{Monthly Payment Toward Loan} = \text{Principal and Interest Balance} \div \text{Number of Months}$$
$$\$60.90 = \$182.70 \div 3$$

Self-Check Answer
$510

Self-Check

Mikovec Electric purchased a copier for $1,500. Find the amount of each payment if it paid for the copier in three equal monthly installments and 12% interest was charged.

Problems

Find the amount due each month on the following installment sales if equal payments are made against the principal. Carry out the interest rate for each month to four decimal places.

1. Donaldson Video Court purchased a video game on the installment plan. The cash price was $159.99 less a down payment of $19.99. The balance was paid off over 4 mo with interest at 10% on the unpaid balance.

   a. Complete this table to determine the balance due each month.

Payment Number	Unpaid Balance	Interest for Month	Amount Due	Installment Payment	Balance Due
1	_____	_____	_____	_____	_____
2	_____	_____	_____	_____	_____
3	_____	_____	_____	_____	_____
4	_____	_____	_____	_____	_____

   b. What would the equal monthly payments for principal and interest be? (Round up your final answer to the next whole cent.)

2. Calorie Counting Cafeteria is replacing its central air conditioner at a cost of $3,500. After a 10% down payment, the balance will be paid in 6 monthly installments, with interest at 9% on the unpaid balance. What is the balance due each month?

Payment Number	Unpaid Balance	Interest for Month	Amount Due	Installment Payment	Balance Due
1	_____	_____	_____	_____	_____
2	_____	_____	_____	_____	_____
3	_____	_____	_____	_____	_____
4	_____	_____	_____	_____	_____
5	_____	_____	_____	_____	_____
6	_____	_____	_____	_____	_____

3. Roland replaced his heavy-duty mower at a price of $627. He was required to pay 25% down and will pay off the balance over 3 mo with interest at 11%. In addition, he was charged $15 for a credit report, which will be included in his monthly payments. What is the balance due each month?

Payment Number	Unpaid Balance	Interest for Month	Other Charges	Amount Due	Installment Payment	Balance Due
1	_____	_____	_____	_____	_____	_____
2	_____	_____	_____	_____	_____	_____
3	_____	_____	_____	_____	_____	_____

4. The replacement cost for a checkout scanner is $1,323.53. The old unit has a trade-in value of $198.53. In addition, there is an insurance charge of $\frac{1}{2}$% of the amount of the loan. There is also a processing charge of $37.12. With interest charged at $9\frac{1}{2}$% for 9 mo, prepare a schedule to show how much is due each month and the balance.

Payment Number	Unpaid Balance	Interest for Month	Other Charges	Amount Due	Installment Payment	Balance Due
1	_____	_____	_____	_____	_____	_____
2	_____	_____	_____	_____	_____	_____
3	_____	_____	_____	_____	_____	_____
4	_____	_____	_____	_____	_____	_____
5	_____	_____	_____	_____	_____	_____
6	_____	_____	_____	_____	_____	_____
7	_____	_____	_____	_____	_____	_____
8	_____	_____	_____	_____	_____	_____
9	_____	_____	_____	_____	_____	_____

Student Success Hints

Test Results. Find the type of problems that you missed, determine if you understood the problem, and determine if any mistakes were careless.

Using a Shortcut for Computing Interest on Unpaid Balances

Rather than computing the interest separately for each payment, you can use the following formula:

$$\text{Monthly payment toward loan} = \text{principal and interest balance} \div \text{number of months}$$

where the interest on the unpaid balances can be found by the formula:

Interest on unpaid balances =

$$\frac{\text{first month's interest} \times (\text{number of payments} + 1)}{2}$$

When the monthly payment is not an exact number, the usual practice is to round up to the next whole cent. This will give a final payment that is slightly smaller than the other payments.

EXAMPLE

Use the formula to find the monthly payment on the lawn mower described in the first example.

First Month's

Interest	= Principal	× Rate	× Time
$1.35	= $180	× 0.09	× $\frac{1}{12}$

Interest on unpaid balances =

$$\frac{\text{first month's interest} \times (\text{number of payments} + 1)}{2}$$

$$= \frac{\$1.35 \times (3 + 1)}{2}$$

$$= \frac{\$5.40}{2} = \$2.70$$

Monthly Payment Toward Loan	=	Principal and Interest Balance	÷	Number of Months

$$\frac{\$182.70}{3} = \$60.90 = (180 + \$2.70) \div 3$$

Self-Check Answer
$521.88

Self-Check
Find the monthly payment if $3,000 is to be repaid in 6 equal installments and the interest rate charged is 15% on the unpaid balance.

5. Paul Graham bought a used delivery van for $4,250. He used his old compact car valued at $250, plus $750 cash, for the down payment. The balance is to be paid in 15 equal installments with interest at 8% on the unpaid balance. Compute (a) the interest on the unpaid balances and (b) the equal monthly payment for the purchase (round up your final answer to the next whole cent).

Answers

5. a. _____

b. _____

Answer

6. _____

6. Peter replaced his old commercial sewing machine with a new computerized model at a price of $2,863. He paid 10% down, and financed the balance at $10\frac{1}{4}\%$ for $2\frac{1}{2}$ years. What is his monthly payment?

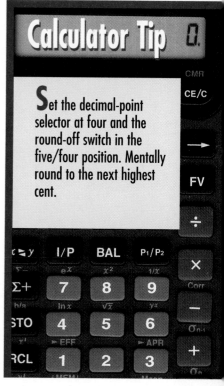

Calculator Tip

Set the decimal-point selector at four and the round-off switch in the five/four position. Mentally round to the next highest cent.

Business Applications 7.8

Answers

1. a. _____

   b. _____

2. _____

3. _____

1. Lorne's Livery purchased a new stretch limousine for $108,000. After trade-in, it obtained a loan for $68,000 at $8\frac{3}{4}\%$ for 7 y. Compute (a) the interest on the unpaid balances and (b) the equal monthly payment for the loan.

2. Herb Neal was able to obtain a $5,200 loan at 7% for 15 mo, to install a new heating system in his restaurant. In addition, he is required to pay $86 insurance, inspection fees of $122, and a credit report charge of $15. What will be his monthly payments?

3. You charge $65.00 on your credit card for dinner. When the bill arrives, you make a $10.00 payment. If your credit card charges 21% interest on the unpaid balance, how much do you own next month?

Activities 7.8

Graph Activity

Kim Tien purchased a washer on May 1 for $390 less a down payment of $90. The balance is to be paid in equal monthly installments over a period of 3 months with interest on the unpaid balance at 12%. Prepare a table of the installment payments and balance due each month.

	Unpaid Balance	Interest for Month	Amount Due	Payment	Balance
June 1	_____	_____	_____	_____	_____
July 1	_____	_____	_____	_____	_____
August 1	_____	_____	_____	_____	_____

A look at statistics that shape your finances

Older women shopping less

Women over age 55 are more apt to say "helpful, friendly" salespeople make shopping easier, while younger women favor hours, parking and price. Outlets with the biggest drop vs. a year ago in female shoppers over 55:

Discount clothing -32%
Specialty clothing -22%
Warehouse clubs -22%
Malls -14%
Department stores -12%

Source: WSL Strategic Retail's How America Shops

By Anne R. Carey and Elys A. McLean, USA TODAY

Challenge Activity

Use a credit card statement or department store charge card statement to see how the interest is charged to the unpaid balance. See if you can arrive at the same interest amount and balance as the credit card company or department store. Write in your own words your findings.

Internet Activity

Credit cards are the most widely used form of credit today. All credit cards charge interest on unpaid balances. Many of the credit card companies also charge interest immediately on cash advances. Compare fees before deciding on a credit card. Check the following web site and learn how to use your credit card wisely.

Go to

http://www.scotiabank.ca/index.html

and select **Scotia Student Link.** Click on **Credit for Students.**

SKILLBUILDER 7.9

Computing Cash Discounts

Learning Objectives

- **Compute cash discounts and net payments.**
- **Compute the balance due after a partial payment.**
- **Determine the cost of credit when a cash discount is involved.**
- **Compute single and multiple trade discounts.**

Computing Cash Discounts and Net Payments

A **cash discount** is a deduction allowed on money owed in order to encourage prompt payment. The rate of discount and the time period allowed to take advantage of the cash discount are called the **terms.** Assume that the terms on an invoice are 3/10, 2/30, n/60. These payment terms provide two opportunities for the buyer to take advantage of a cash discount. The expression 3/10 means that a 3% discount can be taken if the bill is paid within 10 d. The second expression, 2/30, means that only a 2% discount can be taken if payment is made 11 to 30 d after the date of the invoice. More than one cash discount period may be indicated, but only one cash discount is allowed. The expression n/60 means that payment in full is due within 60 d of the invoice date. Cash discounts are never taken on service charges, such as freight, packing, insurance, and so on. If charges of this kind are included in the invoice, they must be deducted before the cash and trade discounts are taken, and then added back in, to find the total amount due.

Cash discount periods are sometimes counted from the end of the month in which the invoice is dated (EOM = end of month) or after the merchandise has been received (ROG = receipt of goods). With EOM, invoices dated from the first day of the month through the twenty-fifth day of the month have a cash discount period from the invoice date to 10 d after the end of that month. If the invoice is dated from the twenty-fifth to the end of the month, then an extra month is allowed for the cash discount period, with the due date being 10 d after the end of the following month. With both EOM and ROG, net payment is due 20 d after the last cash discount date, unless other terms are provided.

MATH TIP

Discount is sometimes given as a percent of the regular price. This is the discount rate. Discount rate × regular price = discount.

EXAMPLE

An invoice for $1,700, dated May 28, has terms of 3/10, 2/30, n/45. What is the last date on which each cash discount can be taken? When is the net payment due? What is the amount due in each instance?

SOLUTION

May 28 + 10 d = June 7

$1,700	Amount of invoice
− 51	Less cash discount: $1,700 × 0.03
$1,649	Net payment

May 28 + 30 days = June 27

$1,700	Amount of invoice
− 34	Less cash discount: $1,700 × 0.02
$1,666	Net payment

May 28 + 45 days = July 12

$1,700	Amount of invoice
− 0	No cash discount allowed after June 27
$1,700	Net payment

EXAMPLE

Terms on an invoice dated August 10 are 3/15, n/45 ROG. The shipment was received on September 29. What is the last date on which the 3% cash discount can be taken, and what is the date for net payment?

SOLUTION

Since the goods were received on September 29, the discount can be taken for 15 d after that date, or until October 14. Net payment is due November 3.

EXAMPLE

An invoice is dated October 30 with terms of 3/10 EOM. What is the last date on which the 3% cash discount can be taken? What is the date for net payment?

SOLUTION

Since the invoice is dated after the twenty-fifth day of the month, we skip November and count 10 d into December for the cash discount period, which thus ends December 10. The net payment date is December 30.

Self-Check

a. The terms on a $2,893.51 invoice are 2/15, n/45. How much are the cash discount and the net payment if the bill is paid within the cash discount period?

b. The terms on an invoice dated May 21 are 2/10, n/30 EOM. What is the last date on which the 2% cash discount can be taken, and what is the date for net payment?

Self-Check Answers

a. $57.87; $2,835.64

b. June 10; June 30

Problems

1. Assume that these bills are paid within the cash discount period. Find the cash discount and the net payment for each invoice.

Amount of Invoice	Terms	Cash Discount	Net Payment
a. $ 875.00	5/20, n/30	_____	_____
b. $1,658.00	3/10, n/30	_____	_____

2. Find the cash discount, if any, and the net payment for these invoices.

Amount of Invoice	Terms	Date of Invoice	Date of Payment	Cash Discount	Net Payment
a. $3,800.00	3/10, n/30	October 13	October 16		
b. 1,725.30	2/10, 1/30, n/60	November 20	December 7	_____	_____
c. 5,167.51	10/10, 3/30, n/60	June 15	July 16	_____	_____
d. 2,950.00	8/10, 4/30, n/60	March 23	April 24	_____	_____

3. Find the discount period, the cash discount, if any, and the net payment for these invoices.

Date of Invoice	Terms	Date Goods Received	Date of Payment	Discount Period	Amount of Invoice	Cash Discount	Net Payment
a. Nov. 21	2/10, n/30 EOM	Nov. 23	Dec. 8	_____	$727.80	_____	_____
b. Sept. 16	3/15, n/30 EOM	Sept. 22	Oct. 14	_____	1,464.00	_____	_____
c. June 6	3/10, n/30 ROG	June 12	June 20	_____	2,748.00	_____	_____
d. Dec. 11	4/10, n/30 ROG	Dec. 18	Dec. 20	_____	482.00	_____	_____
e. Jan. 6	2/10, n/30 EOM	Jan. 20	Feb. 5	_____	4,226.10	_____	_____
f. Aug. 19	1/10, n/30 EOM	Aug. 30	Sept. 11	_____	864.80	_____	_____

4. Jameson Products received a shipment of electronic parts on October 25. The invoice, in the amount of $985.50, was dated October 20. The cash discount terms were 3/10, n/45 EOM.

  a. What is the last day on which a cash discount may be taken?

  b. What amount must be paid then?

Answers

4. a. _____

  b. _____

5. _____

6. _____

5. A shipment invoiced at $3,650 was received by Gable Inc. on May 17. The invoice was dated April 29, with terms of 3/10, n/30 ROG. If payment is made on May 26, how much would Gable Inc. pay?

6. In the previous problem, if the invoiced amount of $3,650 included shipping charges of $112.50, insurance charges of $36.50, and special packing charges of $25, how much would Gable Inc. pay on May 26?

7. Louise Kurtz received a shipment from Boone Electronics in the amount of $12,500. The invoice was dated February 29 with terms of 5/15 EOM. (a) What is the last day on which the cash discount may be taken? (b) How much will she pay? (c) What is the date for net payment?

Computing Balance Due After a Partial Payment

When approved by the seller, the buyer can make payment of only a portion of the invoice and still receive a cash discount on the amount paid. Any remaining balance paid after the cash discount period is paid net. To determine the amount of credit to be allowed for the partial payment, subtract the rate of the cash discount from 100% and divide the amount of the partial payment by the difference.

EXAMPLE

An invoice in the amount of $12,000, dated May 17, has terms of 3/10, n/30. If a partial payment of $5,000 is made on May 27, what amount would be payable on the net date?

SOLUTION

$12,000.00	Amount of invoice
	Cash discount rate = 3%; 100% − 3% = 97%
− 5,154.64	Partial payment: $5,000 ÷ 0.97
$ 6,845.36	Balance due on net date

Self-Check Answer
$9,473.68

Self-Check
Find the balance due on the net date for an invoice in the amount of $20,000 with the term of 5/15, n/45, when a partial payment in the amount of $10,000 is made within the 15-d cash discount period.

8. An invoice in the amount of $7,200 is dated August 12. Terms are 5/10, 3/30, n/60. A payment in the amount of $3,600 is made on August 23. (a) What is the net payment date, and (b) how much is due at that time?

9. Brenda Freidman purchased a new wood-turning lathe for $1,857. The invoice was dated January 12, with terms of 2/20, n/30. Brenda paid $1,000 on February 2. What amount was due on the net date?

10. Henrietta Chen received a shipment in the amount of $28,300. Included in the amount were $782 for freight, $152 for packing, and $78 for insurance. The invoice was dated May 30, with terms of 3/15, 1/30, n/45. Henrietta made a partial payment of $10,000 on June 12 and a further payment of $8,000 on June 29. (a) What is the net date, and (b) how much is due at that time?

11. James Bardhi purchased new lawn furniture off-season for $2,251. The invoice was dated October 27 with terms of 5/15, n/25. James paid $11,000 on November 10. How much did he pay on the net date?

Determining the Cost of Credit

Many businesses have cash-flow problems; although they are otherwise financially solvent, they may not have immediate access to the cash needed to pay current bills, particularly to take advantage of cash discount provisions. A common practice is to obtain a short-term loan to cover the period from the last cash discount date to the net payment date. This practice works because the borrower of a loan is charged interest, on an annual basis, for the period the money is actually used, whereas the cash discount is a flat percent for that specific time period.

EXAMPLE

An invoice in the amount of $5,000 is dated May 1, with terms of 3/10, n/30. In order to take advantage of the cash discount, it is necessary to obtain a 9% loan for the period from the last cash discount date to the net payment date. Will it be profitable to take out such a loan?

SOLUTION

$5,000	Amount of invoice
− 150	Cash discount: $5,000 × 0.03
$4,850	Amount of loan needed

Cost of loan: $4,850 \times 0.09 \times \dfrac{20}{360} = \$24.25$

$150.00	Amount of cash discount
− 24.25	Cost of loan
$125.75	Amount saved

Self-Check Answer
Yes, $131.37 net savings

Self-Check
Myrtle Ellis received a shipment of VCR cassettes invoiced at $8,700. The invoice is dated March 27, with terms of 2/15, n/30. In order to take advantage of the cash discount, she has been quoted a 12% loan rate from her bank. Should Myrtle borrow the money necessary to pay within the cash discount period? How much will it cost her?

12. Fred Hergesheimer received a $4,200 shipment on August 10. The invoice was dated August 3, with terms of 5/10, 2/20, n/45. On August 14 he obtained a loan at $10\frac{1}{2}\%$ to pay for the shipment. (a) How much should Fred borrow? (b) Will he save any money doing so?

13. Darrell Pfeil purchased 850 cases of canned fruit for $16,285. The invoice was dated November 28, with terms of 4/10 EOM. On December 8, he obtained a loan at 9% to pay for the shipment. (a) How much did he borrow, and (b) what was the cost of the loan? (c) What were his savings, if any?

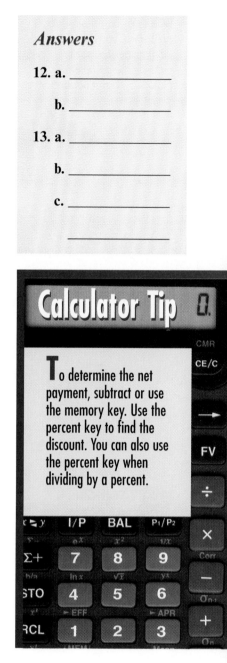

Calculator Tip

To determine the net payment, subtract or use the memory key. Use the percent key to find the discount. You can also use the percent key when dividing by a percent.

Computing Trade Discounts

The price at which an article is sold is the **list price** or **catalog price.** A **trade discount** is a deduction from the list price of an article. A trade discount is offered to qualified purchasers, such as retailers or others who purchase in large quantities. The cost to the retailer after the trade discount is taken is the **net price.**

The discount is expressed as a percent, for example, 15% off the list price. A series of discounts may be offered. When computing the amount of a series of trade discounts, the order in which the discounts are applied will not affect the results.

The net price can also be computed directly. First deduct the trade discount percent from 100% then multiply the list or discounted prices by the remainder.

EXAMPLE

What is the net price of a stereo with a list price of $380, less trade discounts of 10% and 5%?

SOLUTION

First Trade Discount	=	List Price	×	First Trade Discount Rate
$38	=	$380	×	0.10

$380   List price
−   38   First trade discount
$342   Discounted price

Second Trade Discount	=	Discounted Price	×	Second Trade Discount Rate
$17.10	=	$342	×	0.05

$342.00   Discounted price
−   17.10   Second trade discount
$324.90   Net price

To solve using the direct method, subtract each percent from 100%:

$$100\% - 10\% = 90\% \qquad 100\% - 5\% = 95\%$$
$$\$380 \times 0.90 = \$342$$
$$\$342 \times 0.95 = \$324.90$$

Self-Check Answer
$476.43

Self-Check

What is the net price of a compressor with a list price of $590, less trade discounts of 15% and 5%?

14. Find the trade discount and net price of the following.

	Trade Discount	Net Price
a. $50 less 20%	_____	_____
b. $180 less 25% and 15%	_____	_____
c. $900 less 48% and 22%	_____	_____

Student Success Hints

Test Results
Find if the mistakes you made were in the problem-solving steps or the calculations. Did you pace yourself correctly or did time run out before you finished?

15. Use the direct method to compute each net price.

a. $750 less 10%
b. $2,250 less 14% and 20%
c. $400 less 5%, 15%, and 25%

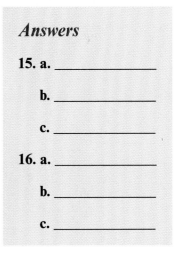

Answers

15. a. _____

b. _____

c. _____

16. a. _____

b. _____

c. _____

16. Use the direct method to compute the list price from the net price in each case.

Net Price	Trade Discount Rate
a. $720	10%
b. $646	5% and 15%
c. $28.56	20%, 30%, and 40%

17. Mary Roman received a shipment of copper tubing on June 2. The invoice was dated May 31, with trade discounts of 15%, 5%, and 10% and terms of 5/10, n/30. Because of a printer error the invoice did not show the list price for the tubing, only the total and a listing of miscellaneous charges. The total for the shipment was $9,376.82. There were charges of $218 for freight, $57 for packing, and $90.12 for insurance. Mary wanted to take advantage of the cash discount period, which had only 8 d to run but wanted to be sure that she received the appropriate trade discounts. What was (a) the list price for the tubing purchased, and what was (b) the final amount which she paid?

18. A cash discount on travel expenses was offered if a companion traveled with the business person. If the original bill is $85 per night for a room, with a 30% discount on the second room, how much do 2 rooms cost each night?

19. Herb Snyder purchased 400 sets of dishes invoiced at $28,500. The invoice is dated April 17 with terms of 4/10, 2/20, n/45. There is also a late payment penalty charge at a rate of 18% per year. Herb made a $10,000 payment on April 26, and a further payment of $10,000 on May 7. Herb did not make the final payment until June 15. What was the amount of the final payment?

Business Applications 7.9

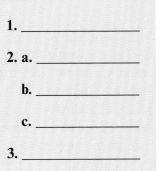

Answers

1. _____

2. a. _____

   b. _____

   c. _____

3. _____

1. Prendevil Products received a shipment invoiced at $5,500, including $300 freight. The invoice is dated April 15, with terms of 4/10,3/30, n/60. Prendevil paid for the shipment on April 29. How much did they pay?

2. Peter's Burgers purchased a deluxe gas grill for $563, including shipping and insurance charges of $31.85. The invoice is dated February 21, 2000 with terms of 3/15, n/45. What is the (a) last date to obtain the cash discount and the (b) date for net payment? (c) If Peter takes the cash discount, how much will he pay for the grill?

3. Melodie Rund received a $7,800 shipment on October 12. The invoice is dated September 29, with terms of 5/15, 3/30, 1/45, n/60. If payment is made on November 14, how much will she pay?

4. Delta Services, Inc. replaced its receptionist's desk and chair with an ergonomically designed set. The desk is list priced at $2,180 with trade discounts of 10% and 5%; the chair is list priced at $875 with trade discounts of 20% and 5%. Sales tax is 6%. There is a delivery and set-up charge of $75. What will be the amount of the final invoice?

5. North Woods Products purchased a total of 125 wood stoves from New England Casters. The order consisted of 40 each Model #32 at $530 list; 40 each Model #38 at $650 list; 30 each Model #42 at $810 list, and 15 each Model #51 at $1,015 list. There are trade discounts of 5%, 7%, and 15%. Model #51 receives an additional 5% trade discount. There is a trucking charge of $20 per unit. Payment terms are 2/10 ROG. The invoice is dated June 15. The shipment was delivered on July 12. In order to take advantage of the cash discount, North Woods took out a loan at 11% on July 21. What was the amount of the loan and its cost?

Activities 7.9

Graph Activity

As the speed of modems increases, the price of lower speed modems left on store shelves decreases. If a $49.00, 33.6 Kbps, modem is discounted by 70%, what would be the new price? _____

Web speed limits

Of the 42% of households with a personal computer, 74% have a modem. Speed of their modem:

Don't know	45%
28.8 Kbps[1]	17%
33.6 Kbps	15%
14.4 Kbps	8%
9.6 Kbps	5%
56 Kbps	5%
Other	4%
12.2 Kbps	1%
T1 or better	1%

1 — Kilobits per second
Note: Does not equal 100% due to rounding.
Source: Maritz AmeriPoll

By Cindy Hall and Web Bryant, USA TODAY

Challenge Activity

Read your bank credit card agreement or your department store credit card agreement to see what terms apply for various situations. You may want to talk to store managers to see what they have to agree to in accepting credit cards in their store.

Internet Activity

Some stores add a surcharge to credit card transactions that are less than $10. Sometimes stores offer cash discounts because they can lose money on a small credit card transaction. Check the following web site and find out what type of person is a responsible user of credit. You may decide to use cash for your next purchase.

http://csf.colorado.edu/lists

Select ESSA, then April 97. Click on Cory Edward Bystrom, Re: Slightly Off Topic

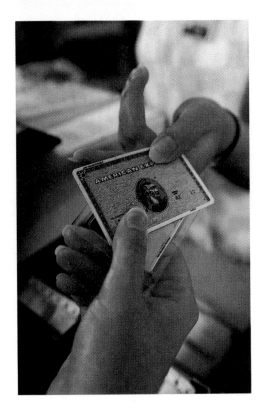

SKILLBUILDER 7.10

Changing a Discount Series to a Single-Discount Equivalent

Learning Objectives

- **Convert a series of discounts to a single-discount equivalent.**
- **Determine the net price directly from the list price.**

Finding the Single-Discount Equivalent

A series of discounts can be converted to a single discount that is equivalent to the discount series. A **single-discount equivalent** is particularly useful when the same series of discounts is used frequently. The single-discount equivalent is *not* the sum of the original discounts because each discount after the first one is based on the **remainder** after the previous discount has been deducted.

EXAMPLE

Determine the single-discount equivalent of the series 20%, 10%, and 5%.

SOLUTION

To find the single-discount equivalent, subtract each discount rate from 100%. Then find the product of these differences.

$$\begin{array}{ccc} 100\% & 100\% & 100\% \\ -\ 20\% & -\ 10\% & -\ 5\% \\ \hline 80\% \times & 90\% \times & 95\% = 68.4\% \end{array}$$

Finally, subtract the result from 100% to get the single-discount equivalent.

$$100\% - 68.4\% = 31.6\% \text{ single-discount equivalent}$$

 Self-Check
Find the single-discount equivalent for a series discount of 10%, 20%, 15%.

MATH TIP

Remember to use the complement of the given discount rate when finding the single-discount equivalent. The complement of a percent is 100% minus that percent.

 **Self-Check Answer**
38.8%

Problems

Find the single-discount equivalent for each series discount. Round decimals to four decimal places, where necessary.

Answers

1. _____
2. _____
3. _____
4. _____
5. _____
6. _____
7. _____
8. _____
9. _____
10. _____

1. 30%, 20%, 10%

2. 25%, 10%

3. 15%, 25%, 5%, 10%

4. $33\frac{1}{3}$%, 15%, 5%

5. 50%, $12\frac{1}{2}$%, $6\frac{1}{4}$%

6. 40%, 20%, 10%

7. 60%, 30%, 15%

8. 15%, $7\frac{1}{2}$%, 5%

9. 40%, 15%, $2\frac{1}{2}$%, 10%

10. 35%, 15%, 5%, 40%, 10%

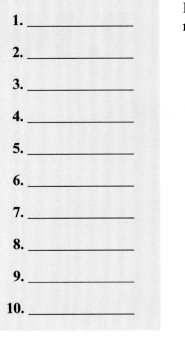

Student Success Hints

Test Results
Draw conclusions from your mistakes on a test to better prepare yourself for future tests.

Finding Net Price

The **net price** is the amount to be charged for an item after all discounts have been applied. To find the net price, the discounts can be subtracted from the list price or a table such as the one shown on page 480 can be used. The horizontal and vertical rate columns list different discounts. Because the order in which discounts are taken does not affect the total discount, there may be several different ways to find the net of $1 (the net price of an article that originally cost $1).

For example, the discount series 10%, 5%, and $2\frac{1}{2}$% can be found in two ways. Read down the left hand rate column until two of the rates in the series are found. If the 5–$2\frac{1}{2}$% row is used, read across this row to the third rate in the series, 10%. The net of $1 for this series is 0.8336. Or, read down further on the left hand column to the 10–$2\frac{1}{2}$% row. Read across this row to the 5% column. The net of $1 is also 0.8336. After the net of $1 is found, multiply it by the total price of the goods purchased to find the net price.

EXAMPLE

An item with a list price of $320 is subject to discounts of 10%, 20%, and $2\frac{1}{2}$%. Find the single-discount equivalent for the item and the net price.

SOLUTION

Read down the rate column until two of the discount rates in the series are found: $10-2\frac{1}{2}$%. Read across this row to the 20% column. The factor at the intersection of this column and row is 0.702, which means that for each dollar subject to discounts of 10%, 20%, and $2\frac{1}{2}$%, only $0.702 would be paid. Multiply this factor (0.702) by the total price ($320). The net price is $224.64. To find the single-discount equivalent, change 0.702 to a percent and subtract it from 100%.

$$100\% - 70.2\% = 29.8\% \text{ Single-discount equivalent}$$

or

$$
\begin{array}{ccc}
100\% & 100\% & 100\% \\
-\ \underline{10\%} & -\ \underline{20\%} & -\ \underline{2\frac{1}{2}\%} \\
90\% \times & 80\% \times & 97\frac{1}{2}\% = 70.2\% \text{ Net factor}
\end{array}
$$

$$100\% - 70.2\% = 29.8\% \text{ Single-discount equivalent}$$

$$\$320 \times 70.2\% = \$224.64 \text{ Net price}$$

or

$$
\begin{array}{l}
\$320.00 \times 10\% = \$32.00 \\
-\ \underline{32.00} \\
\$288.00 \times 20\% = \$57.60 \\
-\ \underline{57.60} \\
\$230.40 \times 2\frac{1}{2}\% = \$5.76 \\
-\ \underline{5.76} \\
\$224.64 \text{ Net price}
\end{array}
$$

Any one of these methods can be used to find the net price when a series of discounts is given.

Self-Check

An item with a list price of $800 is subject to discounts of 20%, 10%, and $7\frac{1}{2}$%.

a. Find the single-discount equivalent for this item.

b. Find the net price of the item.

Self-Check Answers

33.4%

$532.80

Use the table above to find the single-discount equivalent for each discount series.

	Discount Series	*Single-Discount Equivalent*
11.	25%, $7\frac{1}{2}$%	_____
12.	15%, 10%, 5%, $2\frac{1}{2}$%	_____
13.	10%, 40%, 10%, 5%	_____
14.	50%, 25%	_____
15.	10%, 5%, 10%	_____
16.	$7\frac{1}{2}$%, 5%, $7\frac{1}{2}$%	_____
17.	$33\frac{1}{3}$%, 10%, 5%, $2\frac{1}{2}$%	_____
18.	25%, 40%, 5%	_____
19.	40%, 10%, 5%, $2\frac{1}{2}$%	_____
20.	$33\frac{1}{3}$%, 10%, 5%, 25%	_____

Business Applications 7.10

1. Bertram Brothers offered their product subject to discounts of $7\frac{1}{2}\%$, 5%, and 15%. Peach Brothers offer a competing product with discounts of 10%, 5%, $2\frac{1}{2}\%$, and 10%. Both sets of discounts total $27\frac{1}{2}\%$. Who offers the better price?

2. Highlights Unlimited received the following order:

10 doz pens @ $13.58 list; discounts of 30%, 20%, 10%
25 doz writing pads @ $15.87 list; discounts of 50%, 25%, 10%, 5%
4 boxes computer paper @ $49.85 list; discounts of 50%, 20%, 10%
12 packages fax paper @ $49.50 list; discounts of 50%, 10%, 10%, 5%

a. What is the single-discount equivalent for each item?

b. What is the total net price for this order?

3. Ruth Kerlin was billed $299.25 for a fax machine with a list price of $700. The unit was advertised with discounts of 50%, 20%, and 10%. (a) Was Ruth billed correctly? (b) If not, what should the net price have been? (c) What is the single-discount equivalent of the original billing?

4. A leather sofa is list priced at $2,897. The usual retail price is 10% off list. Since this is a commercial purchase, there are trade discounts of 5% and $2\frac{1}{2}$%. The net price of the sofa is 25% above cost. What is the cost price of the sofa?

5. Gracious Gardens Discounters purchased 20 self-propelled mowers at a cost of $159.99 each after trade discounts of 20%, 10%, 5%, $2\frac{1}{2}$%, and 15 mulching mowers at a cost of $143.90 each after trade discounts of 15%, 10%, 10%, 5%. What is the list price of the self-propelled mower, and its selling price after a 20% discount? What is the list price of the mulching mower and its selling price after a 15% discount?

Activities 7.10

Graph Activity

Broadway Series Educational Club bought tickets priced at $80 at a 20% discount and then sold them to college students at 10% off the discounted price. What is the single discount equivalent and how much did the tickets cost each student? _____

USA SNAPSHOTS®

A look at statistics that shape our lives

Sixties good for Broadway

In New York's 1997-98 theater season, 33 new productions opened on Broadway. Seasons since 1959-60 with the most new productions:

NOW PLAYING

1967-68	74
1966-67	69
1965-66	68
1964-65	67
1968-69	67

Source: League of American Theatres and Producers By Anne R. Carey and Gary Visgaitis, USA TODAY

Challenge Activity

Frugal Fran's Fashion Footwear receives trade discounts of 3%, $7\frac{1}{2}$%, $12\frac{1}{2}$%. Fran's selling prices usually range from $7.99 to $39.99. What are the highest and lowest prices she can pay for her inventory? _____

Internet Activity

Imagine that your part-time job is at a bookstore. You are assigned to read *Titan: The Life of John D. Rockerfeller, Sr.* by Ron Chernow. Find this book on the following web site:

http://www.amazon.com

What is the list price? You will receive a 30% discount if you order this book. As an employee, you receive a 25% discount. Today is a special day for employees and you will receive an extra 10% discount. What is your single-discount equivalent? How much will you pay for the book?

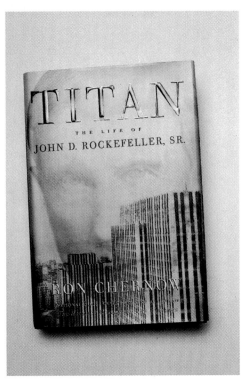

APPENDIX

SINGLE PERSONS — WEEKLY PAYROLL PERIOD

If the wages are—		And the number of withholding allowances claimed is—										
At least	But less than	0	1	2	3	4	5	6	7	8	9	10
		The amount of income tax to be withheld is—										
$0	$55	$0	$0	$0	$0	$0	$0	$0	$0	$0	$0	$0
55	60	1	0	0	0	0	0	0	0	0	0	0
340	350	44	36	29	21	13	5	0	0	0	0	0
350	360	46	38	30	22	14	7	0	0	0	0	0
360	370	47	39	32	24	16	8	0	0	0	0	0
370	380	49	41	33	25	17	10	2	0	0	0	0
380	390	50	42	35	27	19	11	3	0	0	0	0
390	400	52	44	36	28	20	13	5	0	0	0	0
400	410	53	45	38	30	22	14	6	0	0	0	0
410	420	55	47	39	31	23	16	8	0	0	0	0
420	430	56	48	41	33	25	17	9	2	0	0	0
430	440	58	50	42	34	26	19	11	3	0	0	0
440	450	59	51	44	36	28	20	12	5	0	0	0
450	460	61	53	45	37	29	22	14	6	0	0	0
460	470	62	54	47	39	31	23	15	8	0	0	0
470	480	64	56	48	40	32	25	17	9	1	0	0
480	490	65	57	50	42	34	26	18	11	3	0	0
490	500	67	59	51	43	35	28	20	12	4	0	0
500	510	68	60	53	45	37	29	21	14	6	0	0
510	520	70	62	54	46	38	31	23	15	7	0	0
520	530	72	63	56	48	40	32	24	17	9	1	0
530	540	75	65	57	49	41	34	26	18	10	3	0
540	550	78	66	59	51	43	35	27	20	12	4	0
550	560	81	68	60	52	44	37	29	21	13	6	0
560	570	83	69	62	54	46	38	30	23	15	7	0
570	580	86	72	63	55	47	40	32	24	16	9	1
580	590	89	74	65	57	49	41	33	26	18	10	2
590	600	92	77	66	58	50	43	35	27	19	12	4
600	610	95	80	68	60	52	44	36	29	21	13	5
610	620	97	83	69	61	53	46	38	30	22	15	7
620	630	100	86	71	63	55	47	39	32	24	16	8
630	640	103	88	74	64	56	49	41	33	25	18	10
640	650	106	91	77	66	58	50	42	35	27	19	11

MARRIED PERSONS — WEEKLY PAYROLL PERIOD

If the wages are—		And the number of withholding allowances claimed is—										
At least	But less than	0	1	2	3	4	5	6	7	8	9	10
		The amount of income tax to be withheld is—										
$0	$125	$0	$0	$0	$0	$0	$0	$0	$0	$0	$0	$0
125	130	1	0	0	0	0	0	0	0	0	0	0
440	450	48	40	33	25	17	9	1	0	0	0	0
450	460	50	42	34	26	18	11	3	0	0	0	0
460	470	51	43	36	28	20	12	4	0	0	0	0
470	480	53	45	37	29	21	14	6	0	0	0	0
480	490	54	46	39	31	23	15	7	0	0	0	0
490	500	56	48	40	32	24	17	9	1	0	0	0
500	510	57	49	42	34	26	18	10	3	0	0	0
510	520	59	51	43	35	27	20	12	4	0	0	0
520	530	60	52	45	37	29	21	13	6	0	0	0
530	540	62	54	46	38	30	23	15	7	0	0	0
540	550	63	55	48	40	32	24	16	9	1	0	0
550	560	65	57	49	41	33	26	18	10	2	0	0
560	570	66	58	51	43	35	27	19	12	4	0	0
570	580	68	60	52	44	36	29	21	13	5	0	0
580	590	69	61	54	46	38	30	22	15	7	0	0
590	600	71	63	55	47	39	32	24	16	8	1	0
600	610	72	64	57	49	41	33	25	18	10	2	0
610	620	74	66	58	50	42	35	27	19	11	4	0
620	630	75	67	60	52	44	36	28	21	13	5	0
630	640	77	69	61	53	45	38	30	22	14	7	0
640	650	78	70	63	55	47	39	31	24	16	8	0
650	660	80	72	64	56	48	41	33	25	17	10	2
660	670	81	73	66	58	50	42	34	27	19	11	3
670	680	83	75	67	59	51	44	36	28	20	13	5
680	690	84	76	69	61	53	45	37	30	22	14	6
690	700	86	78	70	62	54	47	39	31	23	16	8
700	710	87	79	72	64	56	48	40	33	25	17	9
710	720	89	81	73	65	57	50	42	34	26	19	11
720	730	90	82	75	67	59	51	43	36	28	20	12
730	740	92	84	76	68	60	53	45	37	29	22	14

SIMPLE INTEREST TABLE
$100 ON A 360-DAY BASIS

Time	$2\frac{1}{2}$%	3%	$3\frac{1}{2}$%	4%	$4\frac{1}{2}$%	5%	$5\frac{1}{2}$%	6%	$6\frac{1}{2}$%	7%
1 day	.0069	.0083	.0097	.0111	.0125	.0139	.0153	.0167	.0181	.0194
2 days	.0139	.0167	.0194	.0222	.0250	.0278	.0306	.0333	.0361	.0389
3 days	.0208	.0250	.0292	.0333	.0375	.0417	.0458	.0500	.0542	.0583
4 days	.0278	.0333	.0389	.0444	.0500	.0556	.0611	.0667	.0722	.0778
5 days	.0347	.0417	.0486	.0556	.0625	.0694	.0764	.0833	.0903	.0972
6 days	.0417	.0500	.0583	.0667	.0750	.0833	.0917	.1000	.1083	.1167
7 days	.0486	.0583	.0681	.0778	.0875	.0972	.1069	.1167	.1264	.1361
8 days	.0556	.0667	.0778	.0889	.1000	.1111	.1222	.1333	.1444	.1556
9 days	.0625	.0750	.0875	.1000	.1125	.1250	.1375	.1500	.1625	.1750
10 days	.0694	.0833	.0972	.1111	.1250	.1389	.1528	.1667	.1806	.1944
11 days	.0764	.0917	.1069	.1222	.1375	.1528	.1681	.1833	.1986	.2139
12 days	.0833	.1000	.1167	.1333	.1500	.1667	.1833	.2000	.2167	.2333
13 days	.0903	.1083	.1264	.1444	.1625	.1806	.1986	.2167	.2347	.2528
14 days	.0972	.1167	.1361	.1556	.1750	.1944	.2139	.2333	.2528	.2722
15 days	.1042	.1250	.1458	.1667	.1875	.2083	.2292	.2500	.2708	.2917
16 days	.1111	.1333	.1556	.1778	.2000	.2222	.2444	.2667	.2889	.3111
17 days	.1181	.1417	.1653	.1889	.2125	.2361	.2597	.2833	.3069	.3306
18 days	.1250	.1500	.1750	.2000	.2250	.2500	.2750	.3000	.3250	.3500
19 days	.1319	.1583	.1847	.2111	.2375	.2639	.2903	.3167	.3431	.3694
20 days	.1389	.1667	.1944	.2222	.2500	.2778	.3056	.3333	.3611	.3889
21 days	.1458	.1750	.2042	.2333	.2625	.2917	.3208	.3500	.3792	.4083
22 days	.1528	.1833	.2139	.2444	.2750	.3056	.3361	.3667	.3972	.4278
23 days	.1597	.1917	.2236	.2556	.2875	.3194	.3514	.3833	.4153	.4472
24 days	.1667	.2000	.2333	.2667	.3000	.3333	.3667	.4000	.4333	.4667
25 days	.1736	.2083	.2431	.2778	.3125	.3472	.3819	.4167	.4514	.4861
26 days	.1806	.2167	.2528	.2889	.3250	.3611	.3972	.4333	.4694	.5056
27 days	.1875	.2250	.2625	.3000	.3375	.3750	.4125	.4500	.4875	.5250
28 days	.1944	.2333	.2722	.3111	.3500	.3889	.4278	.4667	.5056	.5444
29 days	.2014	.2417	.2819	.3222	.3625	.4028	.4431	.4833	.5236	.5639
1 month	.2083	.2500	.2917	.3333	.3750	.4167	.4583	.5000	.5417	.5833
2 months	.4167	.5000	.5833	.6667	.7500	.8333	.9167	1.0000	1.0833	1.1667
3 months	.6230	.7500	.8750	1.0000	1.1250	1.2500	1.3750	1.5000	1.6250	1.7500
4 months	.8333	1.0000	1.1667	1.3333	1.5000	1.6667	1.8333	2.0000	2.1667	2.3333
5 months	1.0417	1.2500	1.4583	1.6667	1.8750	2.0833	2.2917	2.5000	2.7083	2.9160
6 months	1.2500	1.5000	1.7500	2.0000	2.2500	2.5000	2.7500	3.0000	3.2500	3.5070

TABLE OF DAYS BETWEEN TWO DATES

Day of Month	January	February	March	April	May	June	July	August	September	October	November	December	Day of Month
1	1	32	60	91	121	152	182	213	244	274	305	335	1
2	2	33	61	92	122	153	183	214	245	275	306	336	2
3	3	34	62	93	123	154	184	215	246	276	307	337	3
4	4	35	63	94	124	155	185	216	247	277	308	338	4
5	5	36	64	95	125	156	186	217	248	278	309	339	5
6	6	37	65	96	126	157	187	218	249	279	310	340	6
7	7	38	66	97	127	158	188	219	250	280	311	341	7
8	8	39	67	98	128	159	189	220	251	281	312	342	8
9	9	40	68	99	129	160	190	221	252	282	313	343	9
10	10	41	69	100	130	161	191	222	253	283	314	344	10
11	11	42	70	101	131	162	192	223	254	284	315	345	11
12	12	43	71	102	132	163	193	224	255	285	316	346	12
13	13	44	72	103	133	164	194	225	256	286	317	347	13
14	14	45	73	104	134	165	195	226	257	287	318	348	14
15	15	46	74	105	135	166	196	227	258	288	319	349	15
16	16	47	75	106	136	167	197	228	259	289	320	350	16
17	17	48	76	107	137	168	198	229	260	290	321	351	17
18	18	49	77	108	138	169	199	230	261	291	322	352	18
19	19	50	78	109	139	170	200	231	262	292	323	353	19
20	20	51	79	110	140	171	201	232	263	293	324	354	20
21	21	52	80	111	141	172	202	233	264	294	325	355	21
22	22	53	81	112	142	173	203	234	265	295	326	356	22
23	23	54	82	113	143	174	204	235	266	296	327	357	23
24	24	55	83	114	144	175	205	236	267	297	328	358	24
25	25	56	84	115	145	176	206	237	268	298	329	359	25
26	26	57	85	116	146	177	207	238	269	299	330	360	26
27	27	58	86	117	147	178	208	239	270	300	331	361	27
28	28	59	87	118	148	179	209	240	271	301	332	362	28
29	29	—	88	119	149	180	210	241	272	302	333	363	29
30	30	—	89	120	150	181	211	242	273	303	334	364	30
31	31	—	90	—	151	—	212	243	—	304	—	365	31

NOTE: For leap year, one day must be added to number of days after February 28.

COMPOUND INTEREST TABLE

Period	$\frac{1}{3}$%	$\frac{1}{2}$%	$\frac{2}{3}$%	$\frac{3}{4}$%	1%	$1\frac{1}{4}$%	$1\frac{1}{2}$%
1	1.003333	1.005000	1.006666	1.007500	1.010000	1.012500	1.015000
2	1.006677	1.010025	1.013377	1.015056	1.020100	1.025156	1.030225
3	1.010033	1.015075	1.020133	1.022669	1.030301	1.037970	1.045678
4	1.013400	1.020150	1.026934	1.030339	1.040604	1.050945	1.061363
5	1.016778	1.025251	1.033780	1.038066	1.051010	1.064082	1.077284
6	1.020167	1.030377	1.040672	1.045852	1.061520	1.077383	1.093443
7	1.023567	1.035529	1.047610	1.053696	1.072135	1.090850	1.109844
8	1.026979	1.040707	1.054594	1.061598	1.082856	1.104486	1.126492
9	1.030403	1.045910	1.061625	1.069560	1.093685	1.118292	1.143389
10	1.033837	1.051140	1.068702	1.077582	1.104622	1.132270	1.160540
11	1.037283	1.056395	1.075827	1.085664	1.115668	1.146424	1.177948
12	1.040741	1.061677	1.082999	1.093806	1.126825	1.160754	1.195618
13	1.044210	1.066986	1.090219	1.102010	1.138093	1.175263	1.213552
14	1.047691	1.072321	1.097487	1.110275	1.149474	1.189954	1.231755
15	1.051183	1.077682	1.104804	1.118602	1.160968	1.204829	1.250232
16	1.054687	1.083071	1.112169	1.126992	1.172578	1.219889	1.268985
17	1.058203	1.088486	1.119584	1.135444	1.184304	1.235138	1.288020
18	1.061730	1.093928	1.127047	1.143960	1.196147	1.250577	1.307340
19	1.065269	1.099398	1.134561	1.152540	1.208108	1.266209	1.326950
20	1.068820	1.104895	1.142125	1.161184	1.220190	1.282037	1.346854
21	1.072383	1.110420	1.149739	1.169893	1.232391	1.298062	1.367057
22	1.075957	1.115972	1.157404	1.178667	1.244715	1.314288	1.387563
23	1.079544	1.121552	1.165120	1.187507	1.257163	1.330717	1.408377
24	1.083142	1.127159	1.172887	1.196413	1.269734	1.347351	1.429502
25	1.086753	1.132795	1.180707	1.205386	1.282431	1.364192	1.450945

COMPOUND INTEREST TABLE
$1 COMPOUNDED DAILY FOR A 365-DAY YEAR

Day	2%	2.5%	2.75%	3%	3.5%
1	1.0000547	1.0000684	1.0000753	1.0000821	1.0000958
2	1.0001094	1.0001368	1.0001506	1.0001646	1.0001916
3	1.0001641	1.0002052	1.0022591	1.0002463	1.0002874
4	1.0002188	1.0002736	1.0003012	1.0003284	1.0003832
5	1.0002735	1.0003420	1.0003765	1.0004105	1.0004790
6	1.0003282	1.0004104	1.0004518	1.0004926	1.0005749
7	1.0003829	1.0004788	1.0005272	1.0005748	1.0006707
8	1.0004376	1.0005473	1.0006025	1.0006569	1.0007666
9	1.0004924	1.0006157	1.0006779	1.0007391	1.0008625
10	1.0005471	1.0006842	1.0007532	1.0008212	1.0009584
11	1.0006018	1.0007526	1.0008286	1.0009034	1.0010543
12	1.0006565	1.0008211	1.0009039	1.0009856	1.0011502
13	1.0007113	1.0008895	1.0009793	1.0010678	1.0012461
14	1.0007660	1.0009580	1.0010547	1.0011500	1.0013420
15	1.0008208	1.0010264	1.0011300	1.0012322	1.0014379
16	1.0008755	1.0010949	1.0012054	1.0013144	1.0015388
17	1.0009302	1.0011634	1.0012808	1.0013966	1.0016298
18	1.0009850	1.0012319	1.0013562	1.0014788	1.0017257
19	1.0010397	1.0013003	1.0014316	1.0015610	1.0018217
20	1.0010945	1.0013688	1.0015070	1.0016432	1.0019177
21	1.0011493	1.0014373	1.0015824	1.0017255	1.0020137
22	1.0012040	1.0015058	1.0016579	1.0018077	1.0021097
23	1.0012588	1.0015743	1.0017333	1.0018899	1.0220571
24	1.0013136	1.0016428	1.0018087	1.0019722	1.0023017
25	1.0013683	1.0017113	1.0018841	1.0020545	1.0023977
26	1.0014231	1.0017799	1.0019596	1.0021367	1.0024937
27	1.0014779	1.0018484	1.0020350	1.0022190	1.0025898
28	1.0015327	1.0019169	1.0021105	1.0023013	1.0026858
29	1.0015874	1.0019854	1.0021859	1.0023836	1.0027819
30	1.0016422	1.0020540	1.0022614	1.0024659	1.0028779
31	1.0016970	1.0021225	1.0023369	1.0025482	1.0029740

DISCOUNT TABLE SHOWING NET OF $1 AFTER DISCOUNTS, SHOWN AT TOP AND SIDE, ARE TAKEN OFF

Rate	5%	$7\frac{1}{2}$%	10%	15%	20%	25%	30%	$33\frac{1}{3}$%	40%	50%
2%	0.931	0.9065	0.882	0.833	0.784	0.735	0.686	0.6534	0.588	0.49
$2\frac{1}{2}$%	0.9263	0.9019	0.8775	0.8288	0.78	0.7313	0.6825	0.65	0.585	0.4875
5%	0.9025	0.8788	0.855	0.8075	0.76	0.7125	0.665	0.6334	0.57	0.475
$5–2\frac{1}{2}$%	0.8799	0.8568	0.8336	0.7873	0.741	0.6947	0.6484	0.6175	0.5558	0.4631
$7\frac{1}{2}$%	0.8788	0.8556	0.8325	0.7863	0.74	0.6938	0.6475	0.6167	0.555	0.4625
$7\frac{1}{2}–5$%	0.8348	0.8128	0.7909	0.7469	0.703	0.6591	0.6151	0.5859	0.5273	0.4394
10%	0.855	0.8325	0.81	0.765	0.72	0.675	0.63	0.60	0.54	0.45
$10–2\frac{1}{2}$%	0.8336	0.8117	0.7898	0.7459	0.702	0.6581	0.6143	0.585	0.5265	0.4388
10–5%	0.8123	0.7909	0.7695	0.7268	0.684	0.6413	0.5985	0.57	0.513	0.4275
$10–5–2\frac{1}{2}$%	0.7919	0.7711	0.7503	0.7086	0.6669	0.6252	0.5835	0.5558	0.5002	0.4168
10–10%	0.7695	0.7493	0.729	0.6885	0.648	0.6075	0.567	0.54	0.486	0.405
10–10–5%	0.7310	0.7118	0.6926	0.6541	0.6156	0.5771	0.5387	0.513	0.4617	0.3848
20–5%	0.722	0.703	0.684	0.646	0.608	0.57	0.532	0.5067	0.456	0.38
20–10%	0.684	0.666	0.648	0.612	0.576	0.54	0.504	0.48	0.432	0.36
25%	0.7125	0.6938	0.675	0.6375	0.60	0.5625	0.5250	0.50	0.45	0.375
25–5%	0.6769	0.6591	0.6413	0.6056	0.57	0.5344	0.4988	0.475	0.4275	0.3563
25–10%	0.6413	0.6244	0.6075	0.5738	0.54	0.5063	0.4725	0.45	0.405	0.3375
25–10–5%	0.6092	0.5932	0.5771	0.5451	0.513	0.4809	0.4489	0.4275	0.3848	0.3206

A

Acceptor, 423
Accountant's method of addition, 37
Accounts, 60
Add-on interest, 443–444
Addends, 11
Addition
　checking, 19
　decimal, 119
　estimating answers in, 42
　fraction, 175–176, 178
　grouping in, 16–17, 35, 37
　horizontal, 27
　mixed number, 175–176, 178
　speed in, 15, 16, 35, 37
　vertical, 28
　whole number, 11
Agent commissions, 273
Alignment, 7
Amount of discount, 419
Annual depreciation, 351
Average inventory, 342–343
Averages, 103

B

Balance, 60
Bank statement reconciliation, 311–313
Banker's year, 371
Bar graphs, 256–257
Base, 227, 233–234
Base 10 system, 3
Broker commissions, 273
Business applications, 60, 81, 99
Buying on time, 439–440

C

Cancellation, 191–192, 201
Cash discounts, 461–463
Catalog price, 469
Check registers, 306
Checkbook stubs, 306
Checking accounts, 305–307
　bank statement reconciliation, 311–313
Circle graphs, 258
Collateral, 431
Collection fee, 431
Columns, alignment of digits in, 7
Commercial year, 371
Commissions
　agent, 273
　broker, 273
　sales, 265–266, 268, 273
　variable, 268
Compound interest
　daily compounding, 413
　formula, 403–404
　periodic rate, 408
　tables, 411
Cost of credit, 468
Cost price
　as inventory value, 341
　markup based on, 279–280
　selling at loss, 289
Credit, cost of, 468
Credit balance, 60
Cross factors, 191–192
Crossfooting, 28

D

Daily compounding, 413
Debit balance, 60
Decimal fractions, 111–112
Decimal numbers
　addition, 119
　division, 137, 139, 151–153
　in fraction addition, 178
　multiplication, 131, 145–148
　place values, 113
　reading, 111–112
　renaming, 163–164, 226
　renaming percents as, 225
　rounding, 115, 139
　subtraction, 125
　writing, 111–112, 113
Decimal system, 3
Decrease problems, 238–239
Deductions, payroll, 325–326
Denominator, lowest common, 159–160, 169
Deposit slips, 305
Deposits in transit, 313
Depreciation
　annual, 351
　double declining balance method, 354–355
　rate of, 351
　straight-line method, 351–352
　sum-of-the-years' digits method, 364–365
　units of production method, 361–362
Difference, 45
Digits
　aligning, 7
　place value of, 113
Discount date, 419
Discounting
　interest-bearing notes, 431–432
　noninterest-bearing drafts, 423–424
　promissory notes, 419–420
Discounts
　cash, 461–463
　discount table, 480
　and net price, 469, 478–479
　single-discount equivalent of series, 477
　trade, 469
Dividend, 87
Division
　business applications, 99
　checking, 93
　decimal, 137, 139, 151–153
　estimation, 95
　fraction, 201
　mixed numbers, 203
　shortcuts, 151–153
　whole number, 87
Divisor, 87
Double declining balance depreciation, 354–355
Drafts, 423
　noninterest-bearing, 423–424
Drawee, 423
Drawer, 423
Due date, 419

E

Estimated life, 351
Estimation
　in addition, 42
　in division, 95

Estimation (cont.)
　in multiplication, 76
　rounding in, 41
　in subtraction, 55
Exact interest, 393
Exact time, 373, 395–396

F

Finance charge, 443–444
First-in, first-out (FIFO) inventory, 344–345
Fractional parts
　determining, 207–208
　to find number of items, 210
　using, 215–216
Fractions
　addition, 175–176, 178
　decimal, 111–112
　division, 201
　improper, 162
　lowest common denominator, 159–160, 169
　multiplication, 191–192
　proper, 159
　renaming, 159, 162, 163–164, 226
　renaming percents as, 225
　subtraction, 183–184

G

Graphs
　bar, 256–257
　circle, 258
　defined, 253
　histograms, 255
　line, 253–254
Gross earnings, 317–318
Grouping, in addition, 16–17, 35, 37

H

Histograms, 255
Horizontal addition, 27
Horizontal subtraction, 51
Hourly rate payroll plans, 317–318

I

Improper fractions, 162
Increase problems, 236–237
Installment purchases
　cost of buying on time, 439–440
　installment purchase price, 451–452
Interest
　compound, 403–404
　defined, 371
　exact, 393
　on installment purchases, 439–440, 451–452
　ordinary, 376–377
　principal, 387–388
　rate, 371–372, 387–388
　real rate of, 443–444
　simple, 371–372, 383–384
　and time, 371–372, 373, 387–388
　on unpaid balances, 451–452, 456
Interest-bearing notes, proceeds on, 431–432
Inventory
　average method, 342–343
　cost basis, 341
　first-in, first-out (FIFO) method, 344–345

© Glencoe/McGraw-Hill

PHOTO CREDITS